The Snowdonia Na

A volume in the New N
of British natural histor
and edited by John Giln
Margaret Davies, and Kenneth Mellanby. The New Naturalist has been described by the *Listener* as 'one of the outstanding feats of publishing since the war', and *The Times Literary Supplement* as 'a series which has set a new standard in natural history books'. Founded in 1945, it now contains more than 50 volumes, of which the following have already appeared in Fontana:

The Highlands and Islands
F. FRASER DARLING & J. MORTON BOYD
The Peak District K. C. EDWARDS
A Natural History of Man in Britain
H. J. FLEURE & M. DAVIES
The Trout W. E. FROST & M. E. BROWN
Wild Flowers J. GILMOUR & M. WALTERS
The Open Sea – Its Natural History
Part I: The World of Plankton ALISTER HARDY
Insect Natural History A. D. IMMS
The Life of the Robin DAVID LACK
Life in Lakes and Rivers
T. T. MACAN & E. B. WORTHINGTON
Climate and the British Scene GORDON MANLEY
Pesticides and Pollution KENNETH MELLANBY
Mountains and Moorlands W. H. PEARSALL
The World of the Soil E. JOHN RUSSELL
Britain's Structure and Scenery L. DUDLEY STAMP
The Sea Shore C. M. YONGE

WILLIAM CONDRY

The Snowdonia National Park

Collins
THE FONTANA NEW NATURALIST

First published in the New Naturalist Series 1966
Second Impression 1967
First issued in Fontana 1969
Second Impression July 1969
Third Impression June 1970
Fourth Impression April 1971
Fifth Impression April 1973

Printed in Great Britain
Collins Clear-Type Press
London and Glasgow

Contents

Plates

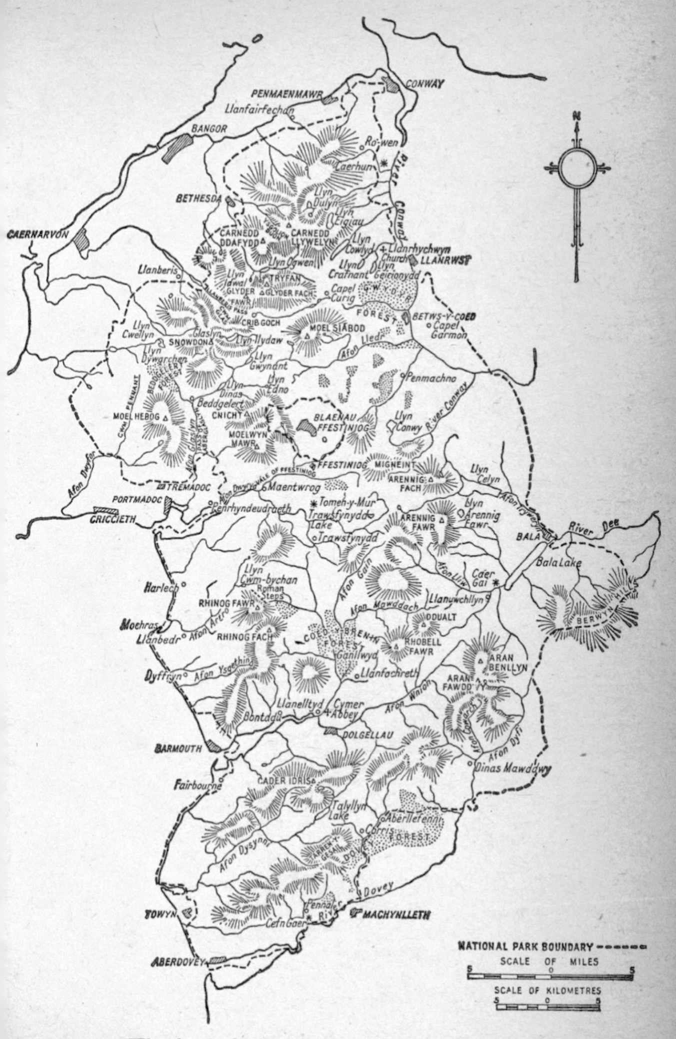

FIG. 1. The boundaries of the Snowdonia National Park

Editors' Preface

The New Naturalist series celebrates, with this book, its coming of age, and a period of remarkable changes in the face of England and Wales, and to an extent also of Scotland and Ireland. The long-overdue planned nature conservation of the kingdom was carried out—indeed, improvised would be a fair word—in the post-war years and has brought us to the situation in which about a tenth of our area is to some extent dedicated to nature, from the giant National Parks to small County Trust Nature Reserves. About a thousand separate areas are now under some form of land management where wild nature has its special acceptance, and hundreds more are sites of special scientific interest, or sites potentially so, under careful watch.

This, then, is the day of the organised naturalist, and of a complex network of initial-bearing bodies ranging from the Natural Environment Research Council and Nature Conservancy—the governmental agencies—and through the Council for Nature to the vast array of private and charitable national and local conservation and nature research societies it serves. Yet though a big and complicated committee structure is now a necessity for wild nature's future, and has evolved and continues to evolve apace (to the extent that, and to the relief of many, it now shows signs of streamlining and simplification), another, older trend shows signs of revival and further evolution.

From the Renaissance to the nineteenth century many naturalists specialised in being general, and were happy to be polymathic. About a century ago came the swing to specialisation: universities and museums needed specialists to bring order to their files and collections, and the majority of zoologists and botanists were happy to embrace more and more authority on perforce fewer and fewer organisms. But with the rise of ecology the all-rounder came back into his own; and with the rise of population pressure and conservation demands he has become a necessity again. Lucky is he who grasps geology, physiography, topography, ecological botany and zoology and—let us add—palaeontology, archaeology and plain history in such depth as lets him see the natural communities of our islands as a whole, as the consequence of an interest in (and the rubbing together of) what were not so long past regarded as many separate disciplines. There is, in short, a new and valuable role for the individual all-rounder, and for an experienced one-man's-eye view of big problems and places. This is why we welcome one such dedicated and knowledgeable naturalist, as writer of this, our new Snowdonian book. William Condry is a private writer, traveller and energetic explorer of the North Wales he lives in and loves; his own master, he is master, too, as our readers will quickly perceive, of the history and natural history of every facet of our second largest National Park.

Snowdonia has been with us as a National Park for fifteen years—long enough to be taken for granted. But it is still something quite peculiar and special—a haven of biologists, archaeologists, palaeontologists, topographers, geologists, historians and climbers that has no parallel in the world. We are convinced that Bill Condry's remarkable book will revive enthusiasm for this great paradise, and that it can feed the new enthusiasts with just the kind of factual meat they will need. It is, indeed, a meaty book, garnished with an elegant style and a nice sense of new values: values which we firmly believe our island civilisation may fully adopt before it is too late, and a second ice age of the Internal Combustion Engine and Infinite Concrete Engineering defeats all the conservationists' efforts.

THE EDITORS

Author's Preface

I have written this book for all who wish to explore the Snowdonia National Park and find out about its scenery, its rocks and soils; its plants, birds and animals. The first half of the book is devoted mainly to chapters on geology and those branches of natural history likely to appeal to most readers. The second half of the book is topographical: I have divided the Park into four sections and attempted to describe them each in turn, pointing out those features of landscape or natural history that have most interested me.

When Linnaeus in the eighteenth century wrote in praise of his native Sweden, Thomas Pennant of Wales was moved to ask: 'Do the heights of Torsburg or Swucku afford more instruction to the naturalist than the mountains of Skiddaw or Snowdon?' The answer is, of course, that they do not. To a naturalist most regions of the earth are good country as long as they are wild and natural enough. Such a region is Snowdonia. The purist may object that there is in fact very little of Snowdonia that is genuinely wild and natural, perhaps only the steepest cliff faces. But compared with much of the rest of southern Britain, Snowdonia is truly wild, as well as being rich and varied. Most of it is either mountain or moorland, considerable tracts of country that have never been disturbed by plough or spade, where it is still possible to pick up a stone axe or flint arrowhead exactly where some hunter dropped it five thousand years ago. Admittedly such a lucky find is improbable, but there is nothing to prevent anyone who has the lungs and the leg-muscles from scrambling up the great terraces of the mountains and finding beautiful and rare alpine flowers that may have been there long before any axe or arrowhead was ever dropped by Neolithic man. But I hope that all who find such flowers will not pick or uproot them but leave them to continue their kind for thousands of years to come. For in the past the Welsh alpine flora has suffered greatly from collecting.

May I offer a word of advice to those keen to understand more about Snowdonia? It is to get to know the people. The real Welsh country people—farmers, shepherds, foresters, quarrymen, roadmen—are often keenly interested in their countryside and are glad to give you all sorts of local facts about rocks and soils, old mines, old roads, old houses and estates; trees, plants (especially herbs), animals and birds; and many other matters which are not written down anywhere but which form valuable basic information for natural history studies. It is well, too, to learn at least a few words of Welsh if only for reading the map intelligently. A vast number of Welsh place-names are descriptive of the landscape and to know what they mean can add greatly to one's appreciation of a place. It is well to remember their antiquity too, for they are among the most ancient words in Britain. Concerning the place-names used in this book: I have followed Welsh spelling except where a place has a long accepted, though barbaric, English form so different from the Welsh that English newcomers to Wales could hardly be expected to recognise them if spelt the Welsh way.

A word of acknowledgment: I have not written about so many places and so many subjects without aid from other people, past and present. Those whose written work has helped me are listed in the bibliography. And of those others who have especially helped me with their advice or information I must mention Peter Benoit, F. C. Best, J. Challinor, A. O. Chater, E. H. Chater, James Fisher, J. Eurfyl Jones, Mary Richards and Evan Roberts. Derek Baylis gave great assistance with the maps of the Park and R. J. Thomas gave valued advice on place-names. I am also much indebted to the staffs of the Nature Conservancy at Bangor and of the National Library at Aberystwyth. I owe much to the late Evan Price Evans whose enthusiasm for mountain ecology was so infectious. Nor must I forget the many pleasurable and profitable days spent in the field with members of the West Wales Naturalists' Trust

(Merioneth Section), for there are few better ways of increasing one's knowledge of a region than by going out with an enthusiastic local society of which every member has something to contribute in the way of facts, comments or useful questions.

Finally, it is well to remember that Snowdonia makes a delightful National Park because it is an area of surpassing beauty and because it contains many habitats rich in fauna and flora. But alas, by no means everyone is sensitive to the need for preserving such things. There are people who would unhesitatingly introduce ugly forms of development and destruction into the heart of a National Park. Those who genuinely care about scenic beauty and wild nature need to be ready to stand up in their defence and make some contribution to the cause of conservation. Apart from the National Parks Commission itself, three bodies exist to uphold this cause: The Council for the Protection of Rural Wales, The North Wales Naturalists' Trust, and the West Wales Naturalists' Trust. All lovers of Snowdonia should support at least one of these excellent organisations.

Introduction

A few words of definition are necessary concerning the name Snowdonia. For centuries it has been virtually a synonym for the Caernarvonshire mountains, especially those immediately around Snowdon itself. But in 1951, the date of the designation of the National Park, the name was suddenly extended to cover a reach of country several times greater in area. As now defined, Snowdonia extends north–south from near Conway to Aberdovey, and east–west from beyond Bala to beyond Tremadoc. It is in this enlarged sense that the word Snowdonia has been used throughout this book.

Having defined its name I can now attempt to define the Park itself with the aid of a few statistics. It is roughly diamond-shaped, some 50 miles from the northern tip to the southern, and about 35 miles from east to west, the area being 845 square miles. The slate-quarrying region of Blaenau Ffestiniog, near the centre of the Park, is excluded. The Park consists largely of blocks of mountain and moorland divided from each other by deep valleys along which most of the main roads have been made. Each mountain block is intersected by many deep radiating valleys. It is the presence of so many valleys so close to each other that is chiefly responsible for the great beauty and variety of the scenery.

The highest peak, which is also the highest point in England and Wales, is Snowdon (3,560 ft.). There are 13 other peaks that top the 3,000-foot contour. All are in Caernarvonshire. Several mountains in south Merioneth almost reach 3,000 feet, the highest being Aran Fawddwy (2,970 ft.) and Cader Idris (2,927 ft.).

The Park has some 22 miles of sea-shore (excluding estuaries), but as the northern boundary falls just short of the coast, the only shore actually in the Park is all in

Merioneth. It consists predominantly of sand and dunes but there are several miles of small cliffs north of Towyn.

Exposed to the predominantly westerly winds of the Atlantic side of Britain, Snowdonia is a mild, wet region, rainfall being greatest in the western half and on the highest ground. Places at low levels close to the west coast have about 40–50 inches average rainfall, but the figure increases rapidly inland and uphill. Dinas Mawddwy has 70 inches, Blaenau Ffestiniog over 90 and Llyn Idwal over 100. The top of Snowdon, reckoned to average 200 inches a year, is one of the four wettest places in Britain, the others being in the Lake District and Scotland. In 1912 a total of 246 inches was recorded on the east side of Snowdon a thousand feet below the summit. On the other hand, absolute droughts are not unknown. On average, spring in Snowdonia is drier than summer, August being particularly wet. But conditions vary greatly from year to year. Generally Snowdonia's climate is mild; but extremely cold conditions can be expected on high ground in easterly winds in winter. From autumn to spring frost and ice are frequent on the mountains and can produce very dangerous conditions on what are normally quite harmless slopes. Normally snow lies on the hills farthest inland, such as the Berwyn range, on more days of the year than on much higher mountains nearer the sea. But there are gullies on Snowdon and the Carneddau where snow may last well into the summer. In 1951 snow persisted in a gully on the north-east side of Carnedd Llywelyn until the last day of July.

CHAPTER ONE

Past Visitors

Since it adds to the interest of a place to see what others have said about it, especially long ago, I have collected in this chapter some of the comments of past visitors to Snowdonia. The number and volume of such writings are vast and the collection of them in the National Library of Wales is surely unique. Reading them, I have marvelled that so many travellers came to Wales in the days when the scarcity of roads and hotels made journeying unbelievably difficult. It is perhaps even more remarkable that so many of these outpourings found a publisher. Did most of these gentlemanly travellers pay for publication, or was there really such a wonderful interest in Snowdonia that the publishers were ready to risk their money on a stream of books about one small region? It was a stream that began about 1770 and overflowed its banks in the 1790's, when a new Tour of Snowdonia appeared on average about every six months for almost the whole decade. Faced with this flood of not always inspiring literature, I have had to be very selective and I will mention only those writers who have particularly impressed me by the freshness of their style, the newness of their observations of people or places, and especially their contributions to natural history.

Neither tourists nor naturalists appear until the seventeenth century. Before then stretch all the centuries of the Dark and Middle Ages from which survives almost no record of what visitors thought or observed about the topography or natural history of Wales. What we know of the Welsh scene in those far-off times—its forests, hunting-laws and game, the state of its agriculture, the way people lived—has been and is being pieced together by the researches of Welsh historians. But as neither this chapter nor this book could hope to encompass so vast a subject I shall have to be content to mention only a handful of commentators from this early period such as Giraldus, Leland, Humphrey Llwyd and Camden.

Giraldus, a native of South Wales, visited North Wales in 1188 in the company of fellow churchmen. Their business was partly with church politics, partly to recruit support for the Third Crusade. Fortunately for us, Giraldus was interested in everything on the way, was gifted with an observant eye and a literary flair, and has left us some invaluable observations. It is fascinating to think of this journey through the Wales of so long ago when all roads must have been green roads or, if you travelled along the coast, there was usually no road at all and you simply followed the line of cliffs and beaches. Giraldus's party came a long way up the coast of Cardigan Bay, where much of their way was along the shore. They crossed estuary after estuary: Dovey, Dysynni, Mawddach, Artro and Dwyryd and so to Bangor. Quite rightly, Giraldus described the Rhinogs as 'the roughest and rudest district in all Wales', for they still are today as any hill-walker knows who has left the tracks and plunged into the ankle-testing riot of heather and rocks that crown the arid ramparts between Llanbedr and Trawsfynydd.

Bird-watchers remember Giraldus with affection because of his claim to have detected a golden oriole in Caernarvonshire only to hear some doubter in the company muttering that it was probably a woodpecker. Times have not changed. To report a golden oriole today to your county bird-recorder is still to invite polite murmurs about how yellow a green woodpecker can sometimes look. But it is a pity about Giraldus's oriole: had he been able to prove its identity it would have been the first record for North Wales by about 700 years.

John Leland seems to have had a marvellous time touring Britain for nearly ten years as 'the King's antiquary'. His job: to visit libraries, monasteries, colleges and churches in order to make an inventory of objects of antiquarian value. Like Giraldus, he noted down a wide range of observations on his journeys. But though he was in Wales for three years (1536–9) we learn little from him about Snowdonia. For him 'Cregeryi Mountaines' were 'horrible with the sighte of bare stones', and he shared the view of his age that there was no point in visiting such forbidding regions. Another century had to pass before there was even a glimmer of a notion that mountains might become a proper study of mankind. But if Leland is almost silent about Snowdonia, he manages, in almost his only sentence on the subject, to give us a vivid picture of the upland peasants' struggle against unkind soils: 'Caernarvonshire about the shore hath reasonable good corne . . . Then more upward be Eryri Hilles, [Snowdon and adjacent heights] and in them ys very little corne, except otes in sum places, and a litle barle, but scantly rye.'

Our next commentator, the historian Humphrey Llwyd, comes about fifty years after Leland. His *History of Cambria* (1584) is depressing for its version of history as a long and scarcely interrupted war. Fortunately, to his history was added an interesting map, and a short 'Description of Wales' by Sir John Price, who includes even a few natural history observations such as that in Bala Lake there is 'a kind of fish called gwyniad which are like to whitings'. But what is really valuable is his reference to deer as plentiful in Caernarvonshire and Merioneth, for this must have been very close to the last time deer could have been common in North Wales. I wish that we could have had in this volume less Humphrey Llwyd and more Sir John Price, for unlike Leland, Price evidently had a real feeling for wild places: 'North Wales hath been a great while the chiefest seat of the last kings of Britain because it was and is the strongest country within this isle; full of high mountains, craggy rocks, great woods and deep valleys, strait and dangerous places, deep and swift rivers as Dovey, which springeth in the hills of Merioneth and runneth through Mawddwy and by Machynlleth and so to the sea at Aberdyfi, dividing North and South Wales asunder.'

Two years after Humphrey Llwyd's *History of Cambria* came the first edition of *Britannia* by William Camden. This, the most complete survey of antiquities since Leland, went through six editions in the author's lifetime and several afterwards. As well as describing antiquities, *Britannia* included some topographical information.

A final comment from the sixteenth century comes in a *Description of Wales* by an unknown author in 1599. He too makes one or two natural history points such as the presence of gwyniad in Bala Lake and the abundance of herrings off the Merioneth coast, two points which will be repeated endlessly by the writers of the next two centuries. He had read Sir John Price and echoes his admiration of mountains, even going so far as to cover them with eternal snows: 'These mountaines may not unfitly be termed the British Alpes, as being the most vaste of all Britaine, and for their steepnesse and cragginesse not unlike to those of Italy, all of them towering up into the aire and encompassing one farre higher than all the rest, peculiarly called Snowdon Hill though the others . . . are by the Welsh termed Craig Eriry as much as Snowy Mountaines . . . For all the yeare long these lye mantelled over with snow hard crusted together.'

It is not until the seventeenth century that any detail of these mountains of Eryri comes into focus for us. For in the first half of that century scientific botany was born and records of the actual localities of alpine plants began to be made, some with enough detail to give us the exciting experience of finding the same species in those same localities three hundred years later. Snowdon has the distinction of being the first British mountain to have been visited by a botanist seriously intent on making a plant list. This pioneer was Thomas Johnson, 'citizen and apothecarye of London'. In his career we can see vividly the change-over from the herbal, medical approach to the purely botanical. For Johnson brought out an edition of Gerard's *Herball* in 1633 and the first ever list of British plants, with no mention of their medicinal worth, in the very next year.

When he published this first list (called *Mercurius Botanicus*) Johnson had not been to Wales. When he did come to Snowdonia in 1639 he made many new finds and these he included in Part Two of his list published in 1641. Considering that he had only one day on Snowdon and that a wet one, and that he was quite new to mountains and could have had little idea of what plants to expect there, Johnson did astonishingly well. From the many good plants he found he obviously had the luck or the botanical instinct to strike one of the richest localities. This Welsh visit was, alas, the very last botanical trip of this pioneer botanist. He had only time to publish his second list before he was involved on the King's side in the Civil War and died of wounds in 1644, aged about 44. His death was a loss to mountain literature as well as to botany for he was the first Snowdon traveller to entertain his readers with accounts of scrambling above horrifying chasms amid tempests and floods of rain. Nineteen years went by before the next notable botanists came into Snowdonia. They were John Ray and Francis Willughby, the two most celebrated English naturalists of their time. On this visit and another four years later (1662) they were able to confirm some of Johnson's finds and add a few new ones.

After Ray and Willughby we come to Edward Lloyd, or Lhuyd, as he often preferred to call himself, a strange genius who in his late teens unaccountably took up botany in an age when botany was scarcely born. Like Johnson and Ray, Lhuyd sought plants in Snowdonia, but unlike them he came summer after summer and so was the first botanist or traveller to get an intimate knowledge of the mountains. In fact very few even since Lhuyd can have known the high ground from Snowdon to Plynlimon as well as he. In the 1680's and '90's Lhuyd made many first British records of flowering plants and ferns, sending packet after packet of specimens to Ray with whom he corresponded for years but never met. When Ray's *Synopsis* of British plants appeared in 1690 it listed many of Lhuyd's finds. The second edition (1696) mentioned also a bulbous plant with rush-like leaves which mystified both Lhuyd and Ray. This *bulbosa alpina juncifolia* Lhuyd had found 'on the highest rocks of Snowdon'. It was still not properly classified or named when both Ray and Lhuyd died, but eventually this most distinguished of Welsh plants received the name *Lloydia serotina* in honour of its finder. Of its several English names perhaps the most appropriate is 'Snowdon lily'.

Lhuyd's botanising was confined to his younger days. Had he concentrated all his talent on plant-study he would undoubtedly have become a very famous botanist. It is typical of his unspecialising century that he soon turned to other interests and became in turn fossil-expert, antiquary and philologist. He achieved distinction in all these fields but, dying before fifty, he did not live to complete the great project of his life, an encyclopedia of Celtic Britain which was to include natural history as a main section, a section which never got further than note-form. After his death these notes went to private collectors and were eventually lost in fires such as the one that burnt down Hafod Uchdryd, Thomas Johnes's Cardiganshire mansion, in 1810. The responsibility for the loss of these precious manuscripts, Welshmen will tell you with emotion, lies squarely on the shoulders of Jesus College, Oxford, which was offered Lhuyd's papers after his death but declined them. What would we not give now to know what Lhuyd had to report on the animals, birds and butterflies known in Wales nearly three centuries ago? But we must be thankful we have some of his botanical records at least. Lhuyd contributed much and was one of the outstanding minds of his age. True he was credulous in a few matters such as his mistaken notions of the origins of fossils, but generally he stands as a tower of common sense and original observation. He was a pioneer in the practice of gathering scientific and other information by means of widely circulated questionnaires. We meet several examples of Lhuyd's good sense in his edition of Camden's *Britannia* of 1695. For instance, Camden and other early writers had declared categorically that there was perpetual snow on Snowdon. Lhuyd answers them with the voice of authority: 'Generally speaking there's no snow here from the end of April to the midst of September . . . It often snows on the tops of these mountains in May and June, but that snow, or rather sleet, melts as fast as it falls.'

The first noteworthy traveller in Wales in the eighteenth century was Daniel Defoe, who came in the 'twenties.

But Defoe had little taste for the wilds. Even in Montgomeryshire, before he had got into Snowdonia, he says he is 'tired with rocks and mountains'. In Merioneth he borders on panic: 'The principal river is the Tovy [Dovey] which rises among the unpassable mountains which range along the centre of this part of Wales and which we call impassable for that even the people themselves called them so; we looked at them indeed with astonishment, for their rugged tops and the immense heights of them. Here among innumerable summits and rising peaks of nameless hills we saw the famous Cader-Idricks [Cader Idris] which some are of opinion is the highest mountain in Britain, another called Rarauvaur [Aran Fawddwy], another called Mowylwynda [Moelwyn Mawr?].'

In 1726 and 1727 Samuel Brewer collected plants in Snowdonia for Dr. J. J. Dillenius, later Professor of Botany at Oxford. 'Trigyfylchau,' whence Lhuyd, then Brewer, reported many finds, is an old name that long puzzled subsequent botanists until it was identified in 1964 by Dr. R. E. Hughes and J. Neill as the Devil's Kitchen cliffs.

In 1732 there came John Loveday, a traveller not afraid of heights. Loveday climbed Cader Idris and only some appalling weather kept him off Snowdon. But he does not enthuse about mountains. He found them depressingly bleak, rocky and barren and nearly came to grief on Cader, only managing to get back to his hotel by the timely aid of the moon, 'with no small difficulty hobbling over ye stones down ye steep mountain'. Still, his *Diary of a Tour* is a pleasing, unsophisticated narrative, unlike many that were to be written around the close of the century. We get a touch, but alas, only a touch, of eighteenth-century wit in John Torbruck's *Collection of Welsh Travels*, 1737. Much better are two lively letters of 1756 by Lord George Lyttleton of Hagley, Worcestershire. Like nearly all travellers, he mentions the gwyniad, or 'whiting', of Bala Lake but he includes another item among the attractions of the place: 'What Bala is most famous for is the beauty of its women and indeed I there saw some of the prettiest girls I ever beheld. The lake produces very fine trout and a fish called whiting, peculiar to itself, and of so delicate a taste that I believe you would prefer the flavour of it to the lips of the fair maids of Bala.'

But it was the Ffestiniog district that really enraptured Lyttleton and he was one of the first to celebrate the beauty of this vale whose praises were to be sung with oppressive repetition by all the Romantic travellers who began to flood in to Snowdonia some twenty years later. Lyttleton employs the word 'romantic', but his style is fairly sober and eighteenth-century, as witness his description of the Vale of Ffestiniog: 'the most perfectly beautiful of all we had seen . . . With the woman one loves, with the friend of one's heart, and a good study of books, one might pass an age here and think it a day. If you have a mind to live long and renew your youth, come and settle at Ffestiniog. Not long ago there died in this neighbourhood an honest Welsh farmer who was 105 years of age. By his first wife he had thirty children, ten by his second, four by his third and seven by two concubines. His youngest son was 81 years younger than his eldest and 800 persons descended from his body attended his funeral.'

Samuel Jackson Pratt, who got to Ffestiniog many years after Lord Lyttleton, chanced to sleep in the house that Lyttleton had lodged in and was able to get from the proprietress this happy memory of the noble writer: 'Aye, he was always scribbling, poor dear gentleman, when he was within doors, and when he was without he ran up and down hills and dales in such a manner, though neither young nor strong, that folks hereabouts thought him a madman; but his *valet de sham* told us he was only a poet, and making a book about us Welsh people and our country; tho' what he could find here worth putting in a printed book, I cannot think, yet he was quite beside himself with joy and often told my husband that we ought to think ourselves very happy, as we lived in Paradise: for that matter we do not live amiss, considering a poor, lone place; we get fish and game of all sorts in plenty, and now and then can shew a joint of meat with anybody.'

As Lyttleton's letters, though written in 1756, were not published till 1781, they cannot be regarded as a pioneer influence on travel literature. For this we must turn to books such as Joseph Cradock's *Letters from Snowdon* (1770), a book which set the pattern of much of what was to follow. For instance, his account of the ascent of Snowdon in the summer of 1769 becomes the prototype for many subsequent descriptions by others: the dawn departure, the somewhat irrelevant quotations from the poets on the way up, and finally the rhapsodies about the view from the top which almost invariably claimed to include not only the rest of Wales and part of the marches but also Ireland and Scotland. 'I doubt', said Cradock, 'whether so extensive a circular prospect is to be seen in any part of the terraqueous globe.' Then, the view having been thoroughly admired and dissected, the reader is treated to lengthy outpourings about the exalted feelings inspired by these ethereal regions in the bosoms of our Romantic adventurers. Jean-Jacques Rousseau may have successfully brought off this ecstatic style in describing the Alps not long before. But Rousseau imitated at perhaps second- or third-hand by unskilled writers is not good to read. How one wishes they had given us less exaltation and more simple observation of what they actually saw about them. Still, we should give Cradock credit for being something of a literary pioneer, for when Cradock was writing *Letters from Snowdon*, Wordsworth was not yet born.

One specimen of the Cradock style will suffice: '. . . Having passed the bridge, how shall I express my feelings! —the dark tremendous precipices, the rapid river roaring over disjointed rocks, black caverns and issuing cataracts— all serve to make this the noblest specimen of the Finely Horrid the eye can possibly behold—the Poet has not described nor the Painter pictured so gloomy a retreat— 'tis the last Approach to the mansion of Pluto through the regions of Despair.' In fact it is nothing of the sort; it is Aberglaslyn Pass. Not that Cradock or any other traveller-writer kept up such raptures all the time. When their heads are cooler they report the passing scene quite competently and tell us something of the villages, the roads, the people and the way of life of the North Wales of two centuries ago. Yet Cradock is sometimes so weary of the road and so contemptuous of everything and everybody along it you wonder why he undertook the tour at all. But he knew how to make the best of the mountain weather. When his ascent of Snowdon from Llyn Cwellyn was delayed by rain, he and his friends were not caught short of ideas for passing the time. They found shelter in 'a small thatched hut at the foot of the mountain near the lake they call Llyn Cychwhechlyn which I leave you to pronounce as well as you are able . . . We were determined to amuse ourselves as well as we could in this dreary situation. For this purpose we sent for a poor blind harper and procured a number of blooming country girls to divert us with their music and dancing.' It is amazing how traveller after traveller in those days found no difficulty in finding 'a poor blind harper', as if there was one laid on in every parish. Or were many of these incidents invented? Alas, when you have read a few of these tours you soon realise how many of their writers lack any trace of conscience when it comes to unacknowledged borrowing from other men's works, even to the extent of describing places they never visited. But are these sins entirely unknown today?

It is during the 1770's with Cradock leading the field that the enjoyment of scenery becomes for the first time a main purpose of travel. One of Cradock's readers was H. P. Wyndham, who produced an account of a Welsh tour he made in 1774, a brief and agreeable narrative whose preface is worth quoting on Welsh travel:

'The author of the following concise Tour has no other view in the publication of it than a desire of introducing his countrymen to consider Wales as an object worthy of attention. The romantic beauties of nature are so singular and extravagant in the principality, particularly in the counties of Merioneth and Caernarvon, that they are scarcely to be conceived by those who have confined their curiosity to other parts of Great Britain.

'Notwithstanding this, the Welsh tour has been hitherto strangely neglected; for while the English roads are crowded with travelling parties of pleasure, the Welsh are so rarely visited that the author did not meet with a single party during his six weeks journey through Wales. [June–July 1774]

'We must account for this from the general prejudice which prevails, that the Welsh roads are impracticable, the inns intolerable, and the people insolent and brutish.

'The writer of these sheets is happy that he is enabled to remove such discouraging difficulties, and assures the reader that in the low, level counties, the turnpikes are excellent; and that the mountain roads are, in most parts, as good as the nature of the country will admit of; that the inns, with a few exceptions, are comfortable; and that the people are universally civil and obliging.'

Judging by the majority opinion of other travellers, Wyndham was right about Welsh hospitality but much too praiseful of the mountain roads. Perhaps he had insufficient experience of them. After all, he climbed no mountains, for this was not essential to the Romantic traveller. Lakes, rivers, waterfalls, beetling crags, forest-embosomed hill-sides, verdant vales: these were the Romantics' special delights and they could all be seen from the valleys. Some felt themselves obliged to go up Cader Idris or Snowdon, but otherwise there was little love of mountaineering. Most mountains you admired from afar, and pressed on to the Vale of Ffestiniog or Aberglaslyn Pass, which explains why Wyndham could not even find one now famous height: 'I made a diligent enquiry through all Caernarvonshire for the Glyder mountain which Gibson has particularly described and which, from its singularity, I much more wished to have seen than the summits of either Plinlimmon or Snowdon. I could however learn no certain intelligence about it.'

The Gibson referred to was Bishop Gibson, who in 1695 had supervised that edition of Camden's *Britannia* to which Edward Lhuyd had contributed the section on Wales.

Wyndham may have found the Welsh roads empty but he seems to have been only just ahead of a crowd who now begin to pour into Wales, if we are to believe what Cradock wrote in his second book, *An Account of Some of the Romantic Parts of North Wales*, which appeared in 1777:

'As every one now who has either traversed a steep mountain or crossed a small channel must write his Tour, it would be almost unpardonable in me to be totally silent, who have visited the most uninhabited regions of North Wales.'

But meanwhile Cradock, Wyndham and all and sundry found themselves overshadowed. A second Lhuyd had arrived, a second great antiquarian-naturalist, and it is fitting that he too should have been a Welshman. Thomas Pennant (1726–98), already the author of *British Zoology*, *A History of British Quadrupeds* and a *Tour in Scotland*, had now turned his attention to his own country and in 1778 produced his first *Tour in Wales*, the work for which he is now best remembered. For in it Pennant has all the attractiveness of the enthusiastic traveller. Despite all difficulties, riding or walking, he got to the remotest places and for the most part his observations are genuinely his own and well described. Pennant sometimes leans on Lhuyd, it is true, but not nearly so much as later writers lean on Pennant, for his *Tour* took precedence over all others, inspired countless imitators and really got people interested in Wales. As H. E. Forrest put it: Pennant 'made known to the average Englishman a country which had hitherto been as unknown as Central Africa'.

Only naturalists can complain about Pennant. For when he toured Wales he was in full antiquarian spate and his remarks on fauna and flora are disappointingly meagre. We can excuse him for saying nothing new about the plants: 'Botany is not within my province', he says, which is fair enough. But he was a leading zoologist and in his *Tour in Scotland* he had gone out of his way to observe the birds; so it is tantalising that when in Snowdonia, almost the only birds he records are ring ouzel, wheatear, great black-backed gull and cormorant, and that he is no more informative about other creatures.

After Pennant, the deluge; especially in the 1790's as the Continent gets embroiled in war and the door to foreign travel is closed. After that there is no alternative but to go exploring less fashionable regions such as the wilder parts of Britain. Meanwhile, the Honourable John Byng had visited Wales in 1784 and again in 1793, though what he wrote was not published until as late at 1934. He is worth reading for his wit and outspokenness. He frankly did not like mountains except to look at from below. Climbing disagreed with him: 'I now don't sleep well at night but am eternally climbing over rocks, descending precipices, &c., and wake at 4 o'clock in the morning, not to sleep again.'

And now the tours become too numerous to describe them all. They vary much in merit and few of them, except those of one or two botanists, add much to Pennant. What is most striking about them is the faithfulness with which the tourists follow each other's tracks as if there were nothing else to see in Wales. They often came by way of Machynlleth or Dinas Mawddwy. They went at least part-way up Cader Idris, mostly by way of Llyn Cau. They traversed Talyllyn Pass and made much of the lake at the top, Llyn y Tri Greyenyn, the Lake of the Three Pebbles and its legend, though this lake is one of the most insignificant in Wales. They passed through or stayed in Dolgellau, many arriving there from Bala. They inspected Cymer Abbey and continued up the Ganllwyd Valley to the three waterfalls of Rhaeadr Du, Pistyll Cain and Rhaeadr Mawddach. Then, some via Harlech, others via Trawsfynydd, they made for Maentwrog, where they lavished more praise on the Vale of Ffestiniog than perhaps anywhere else in Wales. Here other waterfalls were an almost compulsory pilgrimage. From there on to Beddgelert (a few venturing a terrible track over Moelwyn), then Caernarvon, Llanberis and Bangor, somewhere getting in a climb up Snowdon, or, much more rarely, the Glyder. Most then went on to Conway and away home via Chester, but the more enthusiastic also got round to Nant Gwynant and Betws-y-coed.

But we should not be too critical of our travellers. Their powers of walking would put most of us to shame. Take the example of William Hutton, the Birmingham historian. In his old age Hutton made sixteen tours round North Wales, sometimes on horse but mostly on foot. He made his first ascent of Snowdon in 1799 when he was 76. He wrote his *Remarks upon North Wales* when he was 80, and in his eighty-second year walked from Birmingham to Hadrian's Wall and back—a round trip of 600 miles. We have to keep in mind when we think of these early travellers how unknown Wales was, how rare and rough were the roads, how difficult it was to get reliable guides and how poor the available maps were. Even John Evans's map, which came out in 1795 and was such a great improvement on all predecessors, is not really as good as it was made out to be by Arthur Aikin (1797): 'Of this map it is not easy to speak too highly. Every turning of the road, every winding of every rivulet, is laid down with the most scrupulous exactness and the plan of every mountain is given with such minute accuracy, that a person . . . may distinctly trace the course of the primitive, secondary and limestone ridges throughout the whole of North Wales.'

Aikin's *Tour* does not add much to natural history, though he professes a knowledge of the subject. He says the eagle is 'an occasional visitant of the loftiest crags' but produces no evidence for this. Aikin's speciality was rocks and minerals, and though he incorrectly disputed the presence of volcanic rocks on Cader Idris, his book shows the sort of conjecturing that was going on about the nature of rocks just prior to the arrival of the great geologists of the early nineteenth century.

Aikin is memorable as another of the great pedestrians. He did the whole of North Wales on foot, often covering great distances in a day. He was soon followed by another strong-shanked walker, the indefatigable, anecdotal, faintly botanical Reverend Warner. Distinctly more botanical were the Reverends William Bingley and John Evans, both of whom, though separately, made tours of Snowdonia in the summer of 1798. Bingley distinguished himself by his famous scramble after plants up Clogwyn Du'r Arddu on Snowdon, a historic feat, for it is accepted as the first recorded rock-climb in Britain. But both Bingley and Evans owe much to Lhuyd and Pennant that they do not acknowledge and both of them claim to have found plants which in fact were the discoveries of others. Still, whatever his shortcomings, Bingley did undoubtedly find some good plants. And he did react against the adjective-mongers and their hysterical descriptions of Welsh scenery.

The summers of 1796–7–8 were a peak period for traveller-writers. At least eight of them walked or rode through North Wales at that time: Aikin, Bingley, Evans, Hutton, Skrine, Warner, Wigstead and Lord Verulam; and since most of them got their accounts published immediately, they attracted wide public interest. Lured by all the open-mouthed descriptions of mountains and waterfalls, people began to pour into Snowdonia, mostly well-to-do people who could afford to travel about in carriages and indulge in leisurely exploration. By the summer of 1798 we find Bingley reporting: 'There is an inn at almost every respectable town, where post-chaises are kept; but owing to the great numbers who now make this fashionable tour, delays are at times unavoidably occasioned by their being all employed.'

The leading Romantic writers were well represented. Wordsworth climbed Snowdon by moonlight in 1791 and described it in 'The Prelude'. Coleridge walked through North Wales in 1794. De Quincey came in 1802 and, being only a youth, was inclined to rough it and sleep under hedges. Shelley lived at Tremadoc in 1812, giving moral support to his friend Maddocks in the construction of the embankment across the mouth of the Glaslyn river, an action that was the grossest vandalism in the eyes of another English poet then in Wales, Thomas Love Peacock, who protested eloquently against this destruction of a beautiful estuary. Peacock excelled all others in his devotion to Wales and all things Welsh. He bursts with delight at the Vale of Ffestiniog. He goes at midnight to observe moonlight effects at the waterfall of Rhaeadr Du. On the top of Cader he feels 'how happy a man may be with a little money and a sane intellect'. Finally he meets and marries a Welsh girl.

Mountaineering by moonlight was quite in fashion by the beginning of the nineteenth century, the usual aim being to watch the sunrise from the top of Snowdon or Cader Idris, a sight 'which calls forth every sublime emotion of the soul', claims a member of an Anglo-Welsh party that went up Cader. Helped by plenty of food and ale, this party of young men spent a hilarious night among the summit rocks telling strange stories till sunrise. And what stories could a Welshman not tell on the top of Cader Idris at midnight with a full moon above and a pint or two of beer inside him! Among them would surely have been tales of dread and magic and mystery such as have always been associated with Celtic mountains and lakes; or tales about malignant spirits such as the Brenin Llwyd, the Grey King, whose haunt was the highest mountains and whose prey was humanity.

The flocking of more and more people into Snowdonia created in its turn a demand for more detailed information about where to go, how to get there, where to stay and what to see, than the average *Tour* provided. So the modern guide-book came into being. An early one and one of the best was the *Cambrian Traveller's Guide*, which George Nicholson, a printer of Stourport-on-Severn, brought out in 1808. In this book you can see the transition from the travelogue to the guide-book, for though it gives detailed information in the modern guide-book style, it also leans back into the past to collect lengthy descriptive quotations from Pennant, Wyndham, Aikin, Skrine, Warner, Bingley and the rest. This set a pattern that guide-books followed for many years. The *Cambrian Traveller's Guide* was eventually taken over by the author's nephew, Emilius Nicholson, whose revised and corrected third edition of 1840 is, after a century and a quarter, still a really useful book to take with you round Wales.

Most guide-books, though concentrating on scenery and antiquities, included at least some natural history, especially botany. But there were even a few devoted solely to the study of nature. As early as 1805 Turner and Dillwyn's *Botanist's Guide Through England and Wales* had included sections on the rarer plants of Caernarvonshire and Merioneth. Then in 1830 we get an account of the natural history of the parish of Llanrwst, *Faunula Grustensis*, remarkable for being written by a local man, John Williams, when still in his twenties. This was followed by other local natural histories by various authors.

But the fashionable trend, as book-producing techniques improved, was decidedly towards the illustrated book. This gave a new fillip to travel-description, though now the text tended to get subordinated to the many plates of scenery. We get books like T. Roscoe's *Wanderings in North Wales* (1836), illustrated by fifty-one plates. 'Every age has its prevailing fashion', wrote Roscoe, 'and that of the present is assuredly pictorial embellishment in all its forms and branches.'

Meanwhile, in addition to the guide-books, works of a more scholarly nature were dealing with Welsh topography. Two of these I have found almost unfailing sources of information on a wide range of subjects: William Cathrall's *History of North Wales, Volume* II (1828); and Samuel Lewis's *Topographical Dictionary of Wales* (1833). They are both strictly works of reference and are able summaries of what was then existing knowledge about local history, geography and scenery, natural resources, ancient monuments, old roads, agriculture, mining, quarrying, and even valuable snippets of natural history here and there.

Mid-century seems a suitable point at which to conclude this chapter. For a long time roads into Snowdonia had been improving. Even back in 1777 Cradock had said: 'In Wales one has the pleasure of seeing that they are making daily improvements in roads.' All the same, nothing on a big scale was carried out until just after the turn of the century when roads in the modern style began to be built through valleys such as Nant Ffrancon. By the 1830's most of the through valleys had roads, even Llanberis Pass, which in Pennant's day was probably even worse than the Nant Ffrancon road, which he described as a 'dreadful horse-path'. More roads meant more visitors and by 1831 a visitor could write: 'There is no place more public than the higher ground of Eryri during the summer.'

Inevitably, accounts of Snowdonian travel continued to multiply: guides, descriptions, personal narratives. But of all the nineteenth-century tourists George Borrow was deservedly the most widely read. For Borrow was unique. How many Englishmen have ever mastered the Welsh language in order to read the old Welsh literature? And this before ever visiting Wales! This inimitable eccentric who came through Wales in 1854 must surely have inspired more people to visit Wales than any writer since Pennant. Many have had fun following or trying to follow Borrow's actual route, and in doing so have been led into some pretty wild districts. I have two regrets about Borrow: that only a rather small part of *Wild Wales* deals with Snowdonia and that he was in no sense a naturalist. But we cannot hope for everything in one man. It is enough that Borrow had a discerning eye, an intelligent wit and an entertaining pen. After him it is more than ever noticeable what heavy going many of these travel books were. But this did not prevent the demand for them from continuing to increase: for now the Welsh railways were being built, and soon it would be only a few hours from London to once-far-off places like Aberdovey and Harlech. The age of really popular travel had begun.

CHAPTER TWO

Rocks, Land and Forests

Snowdonia is geologically classic ground, for here in the nineteenth century the first scientific investigations of some of the world's oldest rocks were made. But how the history of these rocks was pieced together by the brilliant field work of such pioneers as Adam Sedgwick, R. I. Murchison, and A. C. Ramsay and his colleagues is not within the scope of this book. For those who will take the trouble it is fascinating to read these geologists' own accounts and see how they felt their way through each problem as it arose. To watch them actually at work and to learn something of their methods, their hopes, their disappointments, their mistakes, their triumphs and even their quarrels, provides a living, exciting approach to the geology of Snowdonia. In the century that has passed since those foundation studies, geologists have been able to give increasing attention to detail, and today, though some important questions remain unanswered, it is possible to give a pretty full description of the nature and distribution of the Snowdonian rocks and of their time-sequence.

People down the ages have regarded mountains as symbols of eternity but in modern times it is the impermanence of mountains that has engaged man's attention. Geologists have shown that the mountains of today have been created from the debris of mountains long extinct and that the mountains we know are gradually disappearing, being slowly weathered and eroded and washed into the ocean to provide the material for the mountains of remote future time. We have to visualise the history of the earth as an everlasting drama played out between dry land and ocean, for land and sea have changed places many times. Because of the built-in tensions of the earth's crust, various parts of the ocean bed have risen above sea-level, perhaps as great mountain chains, have been worn away and have sunk again under the sea: it is an infinitely slow series of oscillations that go on continuously and presumably for ever.

To get a clear idea of the general structure of Snowdonia it is best to start outside the Park, in Anglesey, for there the ancient foundation rocks are at the surface. These are the worn stumps of mountains that had been very largely eroded away before the materials which make up the present Snowdonian mountains had begun to be deposited. How thoroughly those very ancient Anglesey rocks were folded can still be seen in the cliffs by South Stack lighthouse, Holyhead. Then, if you cross Anglesey from Holyhead south-east towards Snowdonia, you traverse many miles of these Pre-Cambrian rocks, so called because it was upon them that the Cambrian rocks were laid down. You can follow the Pre-Cambrian rocks across the Menai Strait in the vicinity of Menai Bridge and into the foothills of Snowdonia, where they disappear under Cambrian rocks in the slate-quarry region of Llanberis. In a few miles these Cambrian rocks disappear in their turn under the great mass of rocks of the next system, the Ordovician, of which the bulk of Snowdonia consists. Go right across Snowdonia and you find these Ordovician rocks in their turn dipping down and losing themselves under a cover of the next system of rocks, the Silurian, which form a perimeter round the whole eastern and southern flank of Snowdonia from the estuary of the Conway right round to that of the Dovey. So in a complete traverse of the Park, though we encounter rocks formed over a vast period of time, we meet only with very old rocks, for even the youngest Snowdonian rocks, the Silurian, are classed as Lower Palaeozoic and are over 400 million years old. Incidentally, if one continued on along this line out of Snowdonia south-east as far as Kent, one would encounter ever newer rocks all the way and in 250 miles would have seen about as complete a succession of the earth's rocks

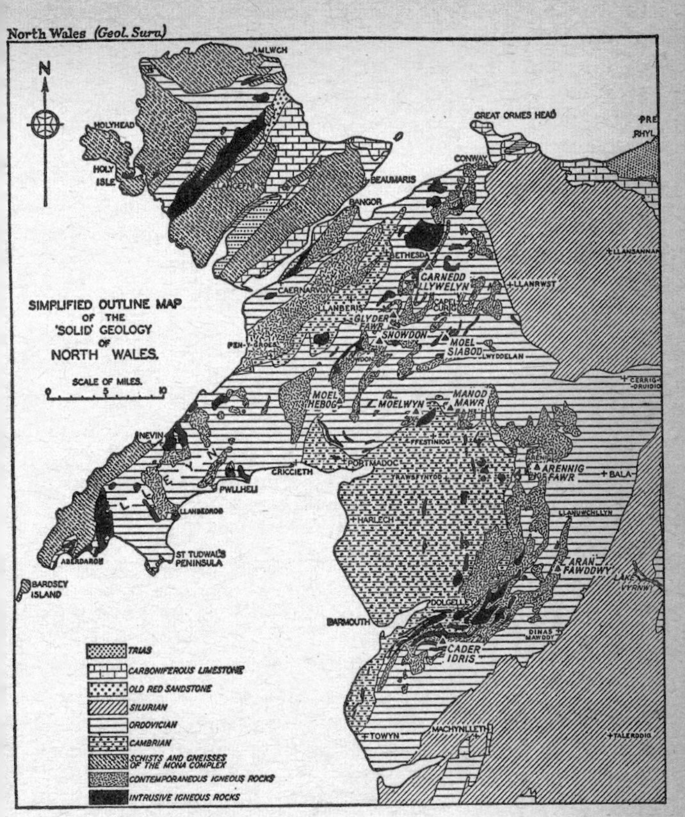

Fig. 2. Geological map of the Snowdonia National Park

from Pre-Cambrian to Recent as could be observed in the same distance anywhere in the world. Such a journey would show how the structure of Snowdonia is related to that of south Britain and would emphasise the ancientness of all the Snowdonian rocks.

Having settled the basic point that the rocks of Snowdonia belong to four very old systems, we can now look at these systems in a little more detail. The Pre-Cambrian rocks do not directly concern us because, although they form the foundation of Snowdonia, they outcrop only outside the Park. In Caernarvonshire, they form a band of acid, igneous rocks stretching between Bangor and Caernarvon and another parallel to it close on the south which, since it includes the north end of Llyn Padarn, is known to geologists as the Padarn Ridge; Pre-Cambrian rocks also extend along parts of the Lleyn Peninsula, including Bardsey Island. Nowhere do they form mountains, for although very hard they are extremely eroded.

The next system of rocks, the Cambrian, was so called by Adam Sedgwick in 1836 because it was first studied by him in 'Cambria', a name for Wales much favoured by Romantic and earlier writers. The Cambrian sediments began to be deposited when the Pre-Cambrian rocks were submerged some 600 million years ago. They are distinguished from the Pre-Cambrian by the presence in them of the first fossil animals—mainly trilobites and brachiopods—whereas the Pre-Cambrian rocks contain only slight and very rudimentary traces of life. The ocean bed of which Snowdonia was then a part evidently sagged continually into a deep trough where west Merioneth is today, for there the Cambrian sediments accumulated to their greatest depth in Britain, 15,000 feet. The wild, barren Rhinog mountains in Merioneth form the largest area of Cambrian rock in Wales. The only other important area of Snowdonia where rocks of Cambrian age are at the surface is in a band, five miles wide (at its widest) and some twenty miles long, which runs along the north-west edge of the Park under the slopes of Snowdon, Glyder and the Carneddau, and so crosses all the three passes that lead from the Menai Strait through the mountains. This band, which runs immediately along the south-east flank of the Pre-Cambrian Padarn Ridge, includes the Nantlle valley, the northern end of Llyn Cwellyn, Llanberis, the high land about Llyn Marchlyn Mawr, and the Bethesda area. This is the Caernarvonshire slate-belt, site of the world's biggest slate-quarries and what is claimed to be the world's finest slate. Here the Cambrian rocks stand up higher than in the Rhinog group and include one three-thousand-footer, Elidir Fawr, in whose flanks the great Dinorwic quarries have been cut, and also Carnedd y Filiast (2,695 ft.) on the other side of the cwm.

Elsewhere, that is nearly throughout Snowdonia, these Cambrian strata lie buried under the rocks of the next system, the Ordovician, which are distinguished from the Cambrian rocks by their fossils. The Ordovician was a period of great volcanic activity. Many volcanic products, such as lavas and ashes, were erupted on to the ocean bed, to be covered subsequently by sedimentary rocks. And massive quantities of igneous rocks—granites, felsites and dolerites—were also intruded into the sedimentary rocks. The name Ordovician commemorates an ancient British tribe who held part of North Wales and were called Ordovices by the Romans. Their Celtic name survives in the place-name Dinorwic—'the fort of the Ordovices'.

The mixture of sediments and volcanic material which had accumulated during Ordovician time remained on the ocean bed for the 30 million years of the next era, the Silurian, whose strata were laid deeply over the Ordovician. These Silurian rocks have now been almost entirely eroded from the National Park area but still cover extensive tracts from the neighbouring Denbighshire moors to central Wales. An important difference between the Ordovician and the Silurian eras is that volcanic activity was negligible during Silurian time. The Silurian deposits, unfortified by igneous rocks, were therefore eroded comparatively easily when later on they were raised above sea-level. The name Silurian was given to this system of rocks in 1835 by Murchison, in honour of the old British tribe, the Silures.

During the Devonian era that followed the Silurian, immense pressures developed in the earth's crust, pressures whose origin is one of geology's unsolved problems. What is certain is that there occurred a succession of earth movements which slowly, over millions of years, uplifted much of Britain, including North Wales, above the sea and folded an immense thickness of rock into long, high, parallel ridges with deep valleys in between, producing a pattern like the waves and troughs of the sea, but with many miles between the top of one wave and the next. One of these ridges arched up over what is now Merioneth, and the bottom of the trough on the north side of this arch was located where Snowdon and its neighbouring peaks now are. Such foldings of the earth's rocks occur when the horizontal advance of part of the crust is resisted by a mass of ultra-hard rocks. When this happens any softer rocks in between get squeezed up and crumpled. In North Wales the pressure came, no one knows why, from the south-east; the resistant block was the very hard Pre-Cambrian rock in the Anglesey region; and the rocks to get squeezed and folded were the relatively less resistant rocks of Merioneth and Caernarvonshire that had been deposited on the ocean beds of Cambrian, Ordovician and Silurian time. Because the pressure came from the south-east, the dominant trend of the folding was from north-east to south-west, that is, at right-angles to the direction of the pressure. The disturbances which caused this folding are known as Caledonian because they also produced the characteristic structure of Scotland, whose north-east to south-west 'grain' is even more pronounced than that of Wales.

But we must not imagine that these pressures produced an orderly pattern of neatly parallel ridges and valleys. The effect was more like that of a choppy sea than of a sea covered with smooth rollers. Just as some parts of a wave rise higher than others, so the great arching of the Merioneth rocks was particularly well developed in the west and has been given the name Harlech Dome. This huge dome is the key to the whole structure of Snowdonia. Not that we are entitled to imagine a dome ever bulging up to say, 20,000 feet and then being eroded. What is more likely is that it was eroded all the time it was being raised. In any case, what is important is the existence of the dome-structure that dictates the positions of the strata of present-day Snowdonian rocks. What is certain is that erosion of the Dome has been very thorough in the Rhinog area. The Silurian and Ordovician rocks have eroded away completely to expose the underlying Cambrian strata (Fig. 3).

Round the Dome, north, east and south, the strata dip beneath harder rocks. It is from these harder rocks that our present-day mountains have been formed. They stand in a semi-circle all round the Rhinog range: Cader Idris, Aran, Arennig, Manod, Moelwyn, Hebog; and then Snowdon, Glyder, Siabod and the Carneddau stretching away to the north-east. They stand up as mountains today chiefly because of the massive quantities of igneous rock associated with them. The sedimentary rocks that covered the igneous rocks have worn away and we are left with our

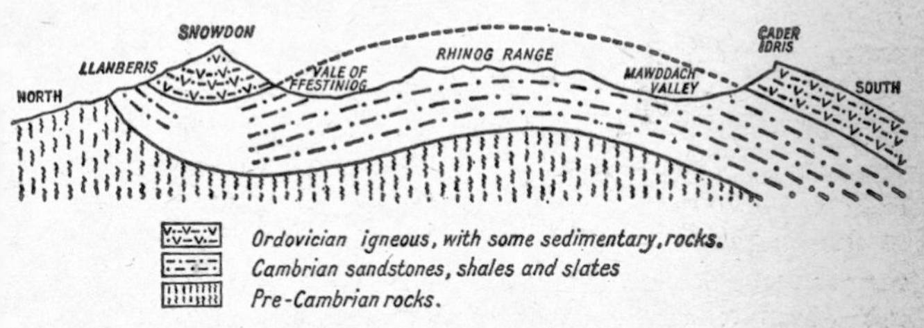

Fig. 3. A generalised section through the Snowdon Syncline and the Harlech Dome

modern mountains, the summits of which consist very largely of igneous rocks. These peaks lie on the map of Snowdonia like a huge figure 5 with the top stroke missing, a figure whose shape was dictated by the distribution of the main blocks of igneous rock (Fig. 4).

The many members of field societies in North Wales who used to have the good fortune to be taken on to Cader Idris by that most enthusiastic of ecologists, the late Price Evans, will recall that it was at this point in his exposition that, like a conjuror, he would produce from his mackintosh pocket a length of thick wire which he proceeded to bend into this beheaded-five shape to give visual aid to his remarks, all the time oblivious of minor inconveniences such as Cader's icy winds or the heavy rain that bounced off his bald cranium. Perhaps his greatest moment was the

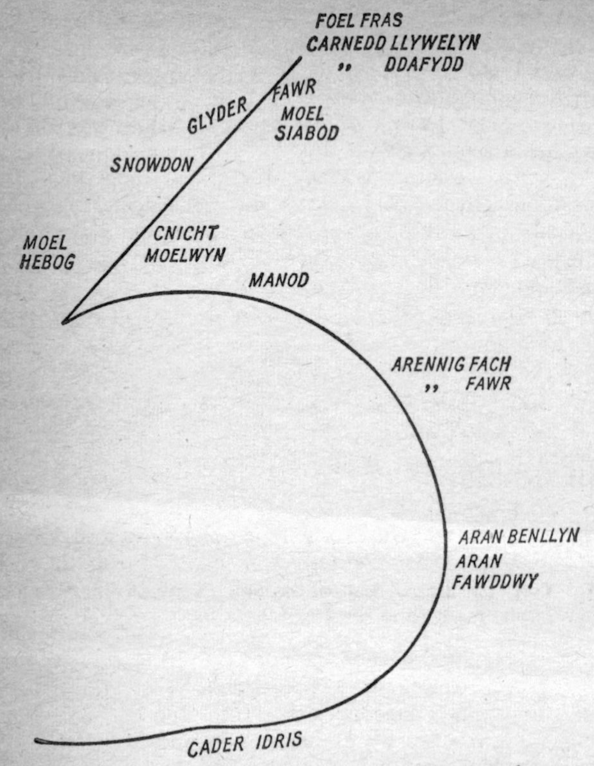

Fig. 4. The igneous rocks of Snowdonia are distributed approximately in the form of a beheaded figure 5

occasion when, in the excitement (and it was a real excitement to him to talk of Cader's rocks), his foot slipped and he fell over backwards; but, holding his figure 5 aloft, he continued to harangue us as he lay on the wet ground rather than interrupt the flow of his thoughts by having to haul his weighty person to his feet again!

It is easy enough to describe the structure of Snowdonia as the remains of a simple arch (anticline) over Merioneth, and a neighbouring trough (syncline) over Caernarvonshire. But it is by no means everywhere that you can actually see the strata illustrating such a structure. You can see them clearly enough in Cwm Cau on Cader Idris, where they dip towards the south and show beautifully the ancient slope of the Harlech Dome. They are visible similarly on the nearby Aran range and on Rhobell Fawr. The bottom of the Caernarvonshire trough (Snowdon syncline) is also clear to see in the downfolded rocks of Clogwyn Du'r Arddu on Snowdon, and the Devil's Kitchen in Cwm Idwal. But over most of the Park the bedding of the rocks is obscured by other divisions in them which to the uninitiated can look exactly like the lines of strata and which are often so pronounced as to conceal the true stratification. These misleading divisions are of two kinds: those caused by jointing and those caused by cleavage. In jointing, a whole series of parallel cracks may develop, often at right-angles to the strata, cutting the rock into rectangular blocks, sometimes of great size. Where these occur in a cliff face they frequently contribute to the formation of those pillars and chimneys so well known among climbers. Jointing results from pressures in the rock and perhaps the weight of overlying strata is sufficient to produce it. But the other sort of splitting, which quarrymen and geologists know as cleavage, is much more fundamental and widespread and has been brought about by the enormous Caledonian pressure of Devonian time. These forces, coming more or less horizontally through the earth's crust, pressed so hard on some of the rocks that their particles came to lie in new positions at right-angles to the direction of the thrust. The result was the formation of new 'planes of cleavage' which were often so completely established that the original bedding of the strata is altogether lost. It is often these planes of cleavage that we see on the faces of slate-quarries and which we must be careful not to mistake for true stratification. But if cleavage is at its highest development in slate, it is present in lesser degrees in most of the other rocks and needs to be kept in mind throughout Snowdonia when we are trying to work out the structure of the landscape.

The pressures which resulted in slaty cleavage had other,

more dramatic effects: that fracturing of the rocks called 'faulting' has also resulted in displacements in the earth's crust so that one side of the fracture has been raised relative to the other or has slipped more or less horizontally—perhaps for some miles—along the plane of fracture. Such great faults can obviously influence scenery in a spectacular way, the best example in Snowdonia being the one extending south-west from Bala where a valley has been formed by erosion along the line of a large fault, the rocks along the fault having been much disturbed and broken up and therefore easily eroded. This valley extends for over thirty beautiful miles from Bala Lake by way of Talyllyn Pass to the coast at Towyn. But lesser faulting occurs throughout Snowdonia and is responsible for many minor scenic effects.

The geological history of Snowdonia in all the vast lapse of time since the Devonian Period is inevitably shadowy, for there are no rocks younger than 400 million years and therefore none of the direct evidence of what has happened such as is afforded by the more recent rocks of an area such as south-east England. The geologist is left to conjecture as best he can how the landscape of today has evolved from that left by the Caledonian mountain-building pressures. He is not even quite sure whether during that time North Wales has been submerged beneath the sea or not; or, if so, how often. The tendency of geological thought is to assume that there have in fact been periods of submergence when rocks later than Silurian were deposited; but if they were they have been entirely removed and there is now only indirect evidence of them, such as is suggested, for instance, by the present drainage pattern. The principal rivers of Snowdonia flow along courses that are unrelated to the fundamental geological structure: the north-east to south-west 'grain'. Little affected by the folds and faults of Devonian time, the rivers have cut through hard and soft rocks alike and often flow across the grain of the country in a manner quite inexplicable unless, as is argued by some geologists, this pattern of our modern rivers was originally cut into a surface of rocks which once lay over Snowdonia but which have now disappeared. Suppose, for instance, that North Wales was submerged under the Cretaceous ocean and received a thick deposit of chalk which domed up over present-day Caernarvonshire. Then such rivers as developed would have radiated out in all directions from roughly where Snowdon is today. These rivers then cut down through the chalk to the Palaeozoic rocks; but by the time they reached them their courses were so well established that they were able to continue them through rocks of varying hardness. So the drainage pattern of one era became super-imposed on the rocks of a previous era. Support is lent to this theory by showing, on a map of present rivers, that by an ingenious linking-up of certain lengths of these rivers with parts of neighbouring rivers, one can produce a pattern of hypothetical rivers which actually do radiate from the vicinity of Snowdon.

So we come to the yesterday of geological time, the past few million years. Such a period looks small on the geologist's time-scale which begins 4,500 million years ago, yet it is probably only in this comparatively recent time that the landscape that we know has been formed. For in that time the modern river pattern has been established. Many rapid and youthful streams have eaten back into the ancient plateau of North Wales, and in doing so have cut into the courses of the old rivers and diverted them down quicker valleys to the coast. Probably, too, a further raising of the land has accelerated the streams and so further increased their cutting power. The result is that the former plateau—the end-result of all the erosion since the Devonian folding took place—has been carved into the pattern of deep valleys and steep-sided mountain blocks of present-day Snowdonia. So it is the modern rivers that have created the modern mountains. As F. J. North well puts it: '. . . the magnificence of our area is due less to the rocks that remain than to those that have been worn away.'

Almost in our own time, geologically speaking, the later details of mountain and valley profiles have been added, largely by the work of the Great Ice Age, in whose tail-end we are living today. The Pleistocene era, which began about a million years ago, is the period of the Great Ice Age, when, between spells of comparative mildness, deep ice covered Snowdonia and most of Britain. Caps of snow, hardened into ice, covered the mountains and from them glaciers radiated down the high slopes and into the valleys. Even at sea-level the summers of that period were powerless to melt the ice, which accumulated on the coast where it met opposing ice-sheets moving down the frozen Irish Sea. Since the last of these glaciers did not finally disappear until 10,000 years ago, it is not surprising that their relics are everywhere abundant and clear to see: the ice-planed pavements and scratched rocks, the many scattered boulders and perched blocks, and the copious moraines. But the boldest effects of the glaciers have been scenic: the hollowing of the mountainsides and the shaping of the valleys. Nearly every mountain in Snowdonia has one flank, sometimes several flanks, deeply rasped out into those dramatic, amphitheatre-shaped hollows which geologists call cirques. Often great cliffs stand round them in a half-circle overhanging a deep, black, moraine-circled lake. Where two cirques have eaten deeply into opposite sides of a mountain, all that may be left on top is a narrow ridge such as the top of Cader Idris between Cwm Cau and Llyn y Gadair. But on higher mountains erosion has often gone a stage further: the two cirques have met and have reduced the summit ridge to a hazardous spine bristling with rough rocks such as are encountered round the Snowdon Horseshoe on Crib Goch and Lliwedd.

The valleys, too, show very pronounced effects of the ice. Unaided, a mountain stream cuts a deep V-shaped valley. But half fill those valleys with an immense weight of slowly moving ice armed with rocks embedded in its underside, and then the valley sides get gouged out and smoothed and the valley's profile changes from V-shaped to U-shaped. Innumerable valleys of Snowdonia show this effect but perhaps none better than the long, straight Nant Ffrancon and the Pass of Llanberis, where you can see also the hanging valleys, similarly U-shaped, which come to an abrupt end high up the sides of the main valleys and are another typical effect of glaciation.

The lakes of the National Park—there are something like 150 of them—are nearly all relics of the Great Ice Age. Either they have been scooped out by glaciers, dammed behind moraines, or both. Cirque-lakes, deep and circular under precipitous cliffs, such as Llyn y Gadair under Cader Idris, Glaslyn under Snowdon, and Llyn Dulyn under Foel Fras, are extreme examples. Enormous concentrations of ice evidently accumulated in these cirques and, heavily charged with rock plucked from the precipices, were able to gouge out these extraordinarily steep-sided basins in the bottom of the cwm. (At one place Llyn Dulyn is 55 feet deep one yard from the shore!) Some geologists have suggested, however, that to excavate such immense holes the glaciers must have cut into rock that had reached a soft and rotten condition—in the way, for instance, that granite in some parts of the world has degenerated into china-clay.

Though practically all of glacial origin, the lakes of Snowdonia vary enormously in character, size, shape, depth and altitude. Not all cirque-lakes are deep: Llyn Idwal, though it reaches 36 feet at one place, is only 10 feet deep over most of its area; and the deepest lake, Cowlyd (222 ft.), has no cirque at its head at all. Nearly all the lakes of the Park are clear and sparkling because of the rocky nature of their surroundings; but a few—shallow saucers on moorlands—are peaty and their waters are therefore reddish-brown. Most lakes are natural; some have been adapted as reservoirs, especially in the Carneddau; a few are wholly artificial, Trawsfynydd and Celyn being among the largest of these. Which are the most beautiful lakes is a matter of choice: some people prefer the wild beauty of Idwal, Llyn Cau, Llyn Hywel, Craiglyn Dyfi and similar lakes where the scenery is essentially water, rock, screes and overhanging crags. Other people favour less savagery, preferring lowland lakes with trees in the picture, more grass on the slopes and more signs of human habitation, lakes such as Gwynant, Crafnant, Talyllyn and, finest and by far the largest of this group, Llyn Tegid at Bala. But all the lakes are beautiful in some way and all are interesting.

If there were no other evidence of it, the presence of so many lakes would indicate a recent period of glaciation in Snowdonia, for if the Great Ice Age were remote in the past, practically all these lakes would have dried up long ago. All lakes are only temporary. Debris brought down by rivers or rolling down mountainsides is always making them shallower; or their outlet streams are ever threatening to drain them by cutting deeper channels. Llyn Ogwen no doubt began life as a fairly deep lake but is now only 10 feet deep at its deepest and will some day become weed-filled, like Llyn Cwm Mynach near Dolgellau; then it will become a quaking morass, such as surrounds the tiny Llyn Crych-y-waen where the Mawddach rises; and finally it will become pasture. We can also prophesy that fairly soon Llyn Ogwen will be cut into two lakes by debris rapidly being brought down the stream from Llyn Bochlwyd which lies in a hanging valley 800 feet above. In this way Llyn Padarn and Llyn Peris were separated from each other by the silt of a mountain stream.

The scenically least dramatic of glacial effects is the part the Great Ice Age played in the production of present-day soils. Glacial drift—the material carried on and within glaciers and deposited by them when they melted—is responsible for vast areas of soil which is therefore unrelated to the rock underneath. But whatever its source, the mountain soil is mostly poor and acid with only relatively small areas of better ground where the underlying rock is richer in minerals. Generally the heavy rainfall has long since carried (leached) the small mineral wealth of the upland soil down from the surface to a depth where it is no longer accessible to most plants. Usually what is removed from the surface by water are compounds of iron which bed down in a tough, impervious. brown layer underneath, leaving the soil above pale, anaemic-looking, and ill-drained. Such a soil, called a podsol (from a Russian peasant-word meaning 'pale soil'), is in Snowdonia often capped with a layer of peat which develops in conditions of acidity and impeded drainage. Such peaty podsols cover miles of mountain and moorland and are especially characterised by a turf containing heather and the mat-grass (*Nardus stricta*), with purple moor-grass (*Molinia caerulea*), cotton-grass and rushes in wetter places.

Such land, apart from its rockiness and often its steepness, is clearly unsuited to any sort of farming except pastoral. So down the centuries great numbers of animals have been kept on the hills: formerly more cattle, ponies and goats, but latterly mostly sheep. The lowest skirts of the hills were formerly much in use for corn-cultivation and in many places today you can still see the outlines of the cornfields of the nineteenth century, often on slopes amazingly steep for ploughland. But comparatively little ground, except in valley bottoms, was ever suitable for corn growing and much corn has always been imported into Snowdonia. In Cathrall's day (early nineteenth century) Caernarvonshire got most of its corn from Anglesey, and as for Merioneth: 'All the corn raised within the county (which is principally rye and oats) is scarcely sufficient to maintain half its inhabitants: and the quarrymen of Ffestiniog are supplied principally from Llanwrst, in Denbighshire, and Evionydd and Lleyn in Caernarvonshire. A great deal of American and other flour has of late years been brought into the ports of Barmouth, Aberdovey and Traeth Bach, near Tremadoc. The best corn land in the country is chiefly along the sea coast, particularly the districts of Ardudwy and Towyn Merioneth; to which may be added the Hundred of Edeirnion, between Bala and Corwen. There is also some good corn land in the parish of Pennal, bordering on the river Dovey.' In our day hardly anyone not directly concerned with their production knows or cares where corn or flour comes from, and home-made bread is now a rarity. There is something depressing in the sight of the hill-farmers, whose ancestors lived and produced their food so independently, motoring down to the village shop to collect the inferior bread turned out by the steam bakeries of distant towns.

The geologist may be interested in many aspects of his science: the age of the rocks; their chemical composition; their physical structure; the architecture of the earth's crust, mineralogy and so on. But what the naturalist most wants to know about rocks is which kinds are associated with which plants and animals, and why. Not only botanists but also some geologists take note of the luxuriance of the vegetation on wet, lime-rich crags compared with its sparseness on drier, acid rocks. For instance, when H. Williams, in 1927, described the volcanic rock in Cwm Glas, he said it was 'unusually calcareous and supports an abundant and typical flora. Indeed in many places the accustomed eye can locate the outcrops of this series with reasonable accuracy by noting the contrast between the patches of heather and bilberry that thrive on the underlying rhyolitic tuffs, and the rich green ferns and mosses that abound on the calcareous pumice-tuffs.' Similarly the lime-rich pillow-lava band in the lower basic rocks of Cader Idris can be traced as clearly by botanists as it can by geologists, because along that band, but not above or below it, grow calcicole (lime-loving) plants such as green spleenwort, bladder fern, purple saxifrage, lesser meadow-rue, mountain sorrel, hairy rock-cress and numerous calcicole mosses. Lichens, living more intimately with rock than other plants do, can be very sensitive indicators of the presence or absence of lime and other minerals, and geologists occasionally make use of them as a quick guide to the nature of rocks. Not that lime itself need be what the calcicole plants are seeking. The importance of lime is that it can make available to the plants certain other minerals that would otherwise remain chemically locked up in the acid soils.

The term 'acid' is unfortunate to the extent that geologists and botanists use it quite differently. Geologists use the term 'acid' in connection with igneous rocks only, which is a very restricted use of the word because plenty of sedimentary rocks are chemically acid too. It is in this wider sense that botanists and agriculturalists use the word 'acid' and in which I have used it throughout this book. Botanists must obviously be careful in reading the geological literature of volcanic regions. The geologists' acid rocks certainly produce the botanists' acid soils. But the geologists' basic rocks need not produce the botanists' base-rich soils; for some of the basic rocks such as certain dolerites can be lime-deficient. In dividing his igneous rocks into acid or basic, what the geologist measures is not the amount of lime in them but the amount of silica. If an igneous rock contains over 65% silica it is an acid rock (granite and rhyolite are typical of this group). If it has between 45% and 52% silica it is a basic rock (such as dolerite and basalt). If it has between 52% and 65% silica it is called intermediate. A rock below 45% silica is termed ultra-basic. Having less silica, the basic rocks are usually richer in magnesium, calcium and iron. Although the geologists' basic volcanic rocks need not be lime-rich, they quite often are, and therefore with them are associated some of the best plant-rich localities. Price Evans's ecological work on Cader Idris brought this out very clearly: the acid volcanic rocks are plant-poor; whereas some of the basic volcanic rocks are plant-rich. Basic volcanic rocks also account for many of the Caernarvonshire plant-rich localities.

Apart from limestone interbedded with volcanic rocks such as on Clogwyn Du'r Arddu, there are narrow bands of it among the acidic sedimentary rocks (slates, grits and mudstones) that cover much of Snowdonia away from the highest peaks. These lime-rich bands have been traced by geologists in several places. In the south, for instance, an interesting narrow band of jet-black shales known only by the Welsh name 'Nod Glas' has been traced all the way from Towyn on the coast, through Abergynolwyn and Corris and on to Dinas Mawddwy, where it turns north to follow the west side of the Dovey valley. For many miles from its starting point at Towyn, the Nod Glas, which varies in thickness from 35 to 70 feet, consists of acid shales; but it becomes associated with a limestone band near Dinas Mawddwy. The presence of this lime profoundly influences the flora of the upper Dovey valley. Not that the Nod Glas or its associated Bala Limestone are visible and continuous bands of rock. On the contrary, they are mostly underground, only exposed where the Dovey's side-streams and, in Llaethnant, the Dovey itself, have cut into them. All the tributaries of the west side of the upper Dovey have a few or many calcicole plants by which botanists could postulate the presence of a lime-rich band even if the geologists were ignorant of it.

The band of limestone along the upper Dovey is evident again to the north of Bwlch y Groes Pass near the farms of Maes Meillion and Gelli Grin, and accounts for the flower-rich meadows thereabouts and onwards into the Hirnant valley. 'Nod Glas' is an old country term which geologists have adopted. The quarrymen of the Corris district have long known this band because it lies at the bottom of the slate-bearing rocks and they came to regard it with dislike for it told them they had reached the end of the slate. Nod Glas means 'blue mark' and seems to have been first named not by quarrymen but by shepherds, who, finding this 'bluish ochre', as Cathrall calls it, where it outcrops on the surface, used it to mark their sheep with. But it is presumably the despair of the quarrymen that is reflected in the Welsh jingle:

O Fawddwy ddu, ni ddaw—dimallan—
A ellir ei rwystraw;
Ond tri pheth helaeth hylaw,
Dyn atgas, nod glas, a gwlaw.

Of this Cathrall's translation is:

In Mawddwy black, three things remain,
False men, blue earth and ceaseless rain:
Of these they'd gladly riddance gain.

The best exposure of the Nod Glas is said to be on the hills between Dinas Mawddwy and Aberllefenni. There the significantly named stream, Nant y Nod, a tributary of Nant Maes y Gamfa, has cut a deep gorge through these soft black shales. It is also exposed not far away at Bwlch Siglen. The Nod Glas is rich in fossils, especially graptolites.

Geologists have also traced bands of limestone through the rocks of Arennig Fawr and its neighbour Moel Llyfnant, but presumably these rocks are not rich in plants for there are few records of calcicoles from that area. These bands of limestone are often rich in fossils. One such band, about two miles south of Arennig Fawr, is named the Ogygia Limestone because it is characterised by the presence of the trilobite (*Ogygia selwyni*) which, says W. G. Fearnsides:

'must sometimes have attained a length of 8 or 9 inches, although more frequently it is represented by specimens only 2 or 2½ inches long, or by numerous tails not more than half an inch across'. Another important band in the Arennig area and one that is full of very interesting fossils (trilobites and brachiopods) is called the Derfel Limestone, from the name of the stream that has cut a gorge across it. Other rich fossil localities are recorded from the upper Llafar valley on Arennig; near Dinas Mawddwy (at Llaethnant and at the junction of the rivers Cowarch and Dovey); and also on the slopes south of Bala Lake. This is in contrast with the nearby mudstones that are practically unfossiliferous over large areas and therefore very frustrating to geologists who would like to put them into chronological sequence. So the acid Ceiswyn mudstones are popular neither with geologists nor naturalists, for their modern fauna and flora are as restricted as their fossils. All the fossil-rich localities in the Park would be too numerous to mention, but—to pick out a few more—there are: the Gwynant valley near Penmaenpool; rocks between the rivers Eden and Mawddach near Dolmelynllyn; rocks near the Snowdon Ranger track above Llyn Cwellyn; Clogwyn Du'r Arddu; the top of Snowdon (the well-known 'fossiliferous ashes' below the hotel); and the Devil's Kitchen. Unfortunately, because of the immense pressures to which the rocks of Snowdonia have been subjected, many of the fossils are distorted and sometimes unidentifiable.

The upland farmers of past centuries knew a lot about the rock among which they lived, for they used it far more than it is used today. They evidently had a nose for so useful a stuff as limestone, for they knew of all its little outcrops, as witness the many old excavations where they dug lime out of the hillsides to burn it and then use it for sweetening their sour land. Perhaps, too, they knew something of how limestone lies in bands through some areas, outcropping at intervals; but it was the early geologists who first clearly grasped this point. So by 1833 the topographer Lewis could report of Merioneth: 'a line of dark-coloured argillaceous limestone extends from Cader

Ddinmael near Cerrig y Druidion, south-westward across the county.' The article then goes on to trace this limestone via Bala (Llwyn y Ci), Llanuwchllyn and Llanfachreth to Cader Idris where, at the farm called Bwlch Coch, a small disused lime-quarry still remains. Lewis also mentions the limestone band extending from the south side of Bala Lake to the upper Dovey at Llanymawddwy, one of the places near which there are traces of ancient excavations for lime, on a hillside called Wenallt (White Slope), a name which may perhaps refer to the lime there.

At what point in prehistory man first learned to find and use metals in Snowdonia is of course conjectural. A widely held belief is that the first metal users in Britain were the beaker-folk who migrated from the Rhine and inaugurated the Bronze Age perhaps 2,000 years before the Romans invaded; but some archæologists would put the use of metal in Britain much earlier. Copper was the first metal widely used and tools of almost pure copper, such as an axe probably from Merioneth, have been found. Then man learned that copper became harder by the addition of tin and it is of the resulting alloy, bronze, that most of the weapons and tools of that age were made: axes, spear-heads, shields, daggers, knives, saws, razors and ornaments. Examples of all these have been found in the Park. There was, for instance, a bronze spear from the scree below Cnicht; a hoard of bronze rapiers from the Rhinog (Cwm Moch); and, finest of all, a bronze circular shield of beautiful craftsmanship found on Siabod and now in the British Museum.

Not that there is any evidence that the copper in these bronze objects was local. The tin certainly was not, for there is no tin in Snowdonia; and if the copper came from Wales it most likely came from the great veins on Parys Mountain, Anglesey, where the copper-mines are known to be ancient. The Romans certainly mined copper in Anglesey and probably also in Caernarvonshire: two copper cakes, stamped in Latin, have been found at the foot of Carnedd Llywelyn. It seems reasonable to suppose that both copper- and lead-mines existed in Snowdonia prior to Roman times and that it was these mines which most attracted the Romans to North Wales. Copper-mines, like other metal-mines in Snowdonia, have always been rather small and scattered because the ore-lodes vary much in richness and are very irregularly and uncertainly distributed in both the Cambrian and Ordovician sedimentary rocks, into which the minerals were precipitated from solutions ascending from the lower heated layers of the earth's crust. Mining has therefore been highly speculative here: fortunes have been made but also lost, and very little metal-mining goes on in Snowdonia these days. In Caernarvonshire copper has been mined at: Llanberis Pass; Ogwen and Nant Ffrancon; Cwm Dyli and other localities on Snowdon; between Beddgelert and Nantmor; the Drws-y-coed and Nantlle area; and in the Conway valley. In Merioneth the copper belt coincides closely with the gold belt and many mines have been worked for both metals. This area extends mainly from Bont Ddu to Rhobell Fawr and includes the Clogau, Glasdir and other mines. Lewis's *Topographical Dictionary* (1833) picks out Benglog on Rhobell Fawr as a mine where 'vast quantities of copper were produced some years ago'. But the most interesting copper in that area was obtained from peat, as Lewis relates: 'A gentleman resident in Dolgelly, learning that peat from Dolfrwynog in the parish of Llanfachreth was useless as a manure, had it analysed and found it contained copper. So he had a great quantity of peat cut and burned in kilns and shipped the ashes to Swansea where they were made to yield excellent copper. From this circumstance it has been supposed that the surrounding mountains teem with copper ore which, through the medium of springs, had impregnated the peat of the hollows below with a solution of sulphate of copper.'

Although the Romans mined for gold in Carmarthenshire, there is no indication that gold-mining in Snowdonia is anything but recent, beginning in the 1840's. The gold-yield has always been small and if there are any rich lodes they are deep in the rocks and would be very difficult to find. Not that people have not tried: there have been innumerable trial-holes made along the middle reaches of the Mawddach, around Bont Ddu and even south of the estuary as far as Fairbourne. Panning in the streams also produces a little dust; and speculators have found traces of gold in the silt of the estuary. One of the largest mines (now derelict) is the Gwynfynydd, where the River Cain flows into the Mawddach. There is also copper thereabouts, but in the seventeenth century Gwynfynydd was known for its lead, for there, as we learn from the reply to Lhuyd's questionnaire, 'they dig very good leaden oar'. There have also been gold-mines three or four miles north-east of Trawsfynydd and near Llanuwchllyn. At present gold is worked (fitfully) only near Bont Ddu. All Snowdonian gold is found in veins of quartz and the mines can easily be identified by the white crushed quartz of their tips. Bronze Age man made much use of gold ornaments, two fine examples of which have been found in Snowdonia. They were both necklaces of twisted gold about four feet long and weighing half a pound. The first was dug up in 1692 in a garden close to Harlech Castle; the second was found in 1823 on the lower slopes of Cader Idris above Llyn Gwernan. But the gold of those far-off times came not from Wales but from the prolific mines of County Wicklow across the Irish Sea.

Except in the lower Conway valley, lead has been of less importance than copper in Snowdonia for the main lodes of Welsh lead lie elsewhere. Nearly all the lead-mines have been derelict since the First World War or earlier, but the Parc mine near Llanrwst is modern and can go into production whenever the market price of lead makes it worth while. The fortunes of the Welsh lead-mines have fluctuated violently for centuries. Their last heyday was the 1860's and 1870's. Then about 1875 larger and more easily worked deposits were discovered in America and elsewhere and Welsh lead was quickly priced out of the market.

Derelict mines can be ugly places and dangerous with tottering walls and unfenced shafts; but their plants and animals can be of great interest. Although copper and lead are poisonous, and have killed many fish in lakes and rivers near mines, the associated non-economic (gangue) minerals brought up incidentally out of the earth along with these metals may prove attractive to certain plants. So at Hermon copper-mine, five miles north-north-east of Dolgellau, there grows an abundance of thrift and vernal sandwort; and in the Llanberis Pass Evan Roberts has also found vernal sandwort near old copper levels. This sandwort is much better known as a lead-mine not a copper-mine plant elsewhere, and is in fact called leadwort locally in Yorkshire. Since it is very much at home on lime-rich crags it is presumably attracted to mines by the presence of calcium carbonate in the gangue. This source of lime may also account for the fairy flax on the quartz tips of the gold-mine near Llanuwchllyn. In abundance in the lower Conway valley is another typical lead-mine plant, alpine penny-cress. Near it, in much smaller quantities, is the forked spleenwort. If the ecology of all these plants is problematical, that of the forked spleenwort is particularly difficult to understand. Its most natural Welsh habitats are hard rocks, which often appear to be acid rocks and in any case are quite different from the softer base-rich rocks on which grow most small ferns such as bladder fern, woodsia, holly fern and green spleenwort. Yet it also flourishes in the loose scree of lead-mine tips; and in Cardiganshire it thrives best in the mortared walls of lead-mines and there looks a distinct calcicole. But if it is a calcicole why is it not found in the usual communities of calcicoles on the mountain ledges?

Wall-nesting birds such as stock doves and wagtails are frequent at lead-mine ruins, and occasionally kestrels and ring ouzels. Bats are rare in upland mines. They are found mainly in lowland districts, spending the day in roofs, especially of churches, large old houses and farm buildings, and also in ivied trees. The only bats I have seen in wilder, semi-upland habitats have emerged from narrow cracks in rocks. But the distribution and habitats of Snowdonian bats are little known and their investigation, admittedly not easy, awaits some patient naturalist or better still a team of naturalists with the time and enthusiasm to go a-batting in belfries, barns, hollow trees, roofs, caves and old mine-workings. Eight species are recorded for the Park: long-eared, pipistrelle, noctule, whiskered, Daubenton's, Natterer's, greater horseshoe and lesser horseshoe. Quite a few bats have been found in North Wales mines outside Snowdonia but the only one I know of as recorded from a mine within the Park is the greater horseshoe, found in a mine near Penmaenpool as long ago as 1896. It must be said that the naturalists of 60 or 70 years ago took more interest in bats than we do today and that we still rely heavily on Forrest's *Fauna* of 1907 for our knowledge of these elusive creatures.

Manganese is chiefly associated with the rocks of Cambrian age that form the Rhinog range and, the Rhinog being the rough place it is, some of the mines lie in difficult rock and heather country such as that between Llyn Cwm Mynach and Diffwys. There are thought to be vast quantities of low-grade manganese ore in the Rhinog area which may be worked in the future to supply the needs of blast furnaces, but none is worked at present. Nor has any iron been produced in Snowdonia for many years. Small iron-mines have been operated near Betws Garmon, Tremadoc, Penrhyndeudraeth and at Tir Stent on Cader Idris.

Easily the most important present-day mineral industry of Snowdonia is the slate industry, the largest quarries being outside the Park boundary at Nantlle, Llanberis, Bethesda, Blaenau Ffestiniog and Corris. Some slate is quarried, some is mined, but however it is won, its getting produces vast tips of waste. This is because only the best slate is saleable and much inferior rock has to be moved aside to get at the marketable slate. The word slate refers to no particular type of rock. It can be used for any rock which splits easily into smooth thin sheets, most of such rocks in Snowdonia having originally been mudstones and shales which, as explained earlier, have been subjected to enormous lateral pressure. Such very altered rocks belong to the major group of rocks which geologists call metamorphic.

It is not only the sedimentary rocks of North Wales that are exploited. There are also large quarries for extracting the intrusive igneous rocks (dolerite and granite). They too are mostly outside the Park, or nearly so, at Penmaenmawr, Blaenau Ffestiniog, Minffordd, Tonfanau and Arennig. Some of the rock, which serves mostly as roadstone, is used locally but the bulk of it goes to England.

'The Country is mountainous and yields pretty handsome clambering for goats and hath variety of precipice to break one's neck; which a man may do sooner than fill his belly, the soil being barren and an excellent place to breed a famine in.' So commented John Torbruck in his *Journey Through Wales* (1749). He can be said to speak for nearly all the early travellers: most of those who condescended to notice that Snowdonia was in fact inhabited were appalled by the poverty of the people and of their cottages. But times have changed, agriculture has been revolutionised, and now the visitor sees a very different peasantry. When you meet Dai Jones on the mountain with his cowering, wall-eyed dogs and you note his three-day beard, his torn old cap and his ancient jacket pinned together with a couple of four-inch nails, you need not pity his low standard of living. In town on market days he looks smart enough and he is not in any sense a poor man. His life may appear rough and primitive but it is a more real life than most of us live. It is certainly a different one, a life in which mountains, rocks, streams, steep slopes, rain, wind, snow, sheep, foxes, ravens and buzzards are the common ingredients. But then he is a different man from most of us. Something of his outlook, like his blood-group, may go straight back to the Celtic people who defended themselves in the hill-forts of the first centuries A.D. and who were probably never subdued by the Romans but co-existed with them and perhaps even drove hard bargains with them over the price of lead and copper. In the life of Dai Jones, in his farm up in some remote side-valley, something of that independence endures today.

The upland pastures may be poor but if they can support but one ewe to an acre (and most do better) that is still a lot of sheep on a whole mountain. And sheep are profitable. So, except where there are conifers, sheep are everywhere on the mountains. Formerly beef cattle grazed up to the highest ground. But when sheep came into fashion, walls were built along the mountainsides dividing sheep above from cattle below. A hill farm may have many walls but 'the mountain wall' is distinct: it is the highest one, above which 'the mountain', as the farmer calls the highest pastures, stretches away to the summits.

The 'wild' goats of Snowdonia, descendants of former domesticated stock, are an interesting relic of the farming scene of a few centuries ago. They inhabit rocky mountainsides, and although they compete with sheep for food, they are generally looked on with favour by the shepherd because, so much more agile than sheep, they graze on the most dangerous ledges and so remove pasture which might otherwise have tempted a sheep to its death. So many of the early travellers remark on the goat that it was evidently a tourist attraction, goats posed along rocky skylines being highly picturesque. But how wasteful and destructive those armies of goats must have been! Fortunately they went out of favour during the eighteenth century because they did so much damage in the young forests that many landowners were trying to establish. Besides, bushy wigs that had been made of goat hair went out of vogue at that time. But goats disappeared only slowly. As late as 1774, while boating on Llanberis lake, Dr. Samuel Johnson solemnly witnessed that Miss Thrale (aged 10) counted 149 goats on the slopes of Snowdon. (Her father had promised her a penny for every one she saw, Dr. Johnson being the referee.) By 1803 William Hutton marks their passing: 'This old and once numerous inhabitant of Wales, like the language, is declining; and like that, will come to a period.' But that period, after more than a century and a half, has not yet come. Some magnificent goats, wild and ibex-like with their towering, swept-back horns, still thrive on the rocks of Tryfan, Rhinog, Rhobell Fawr, Cader Idris and perhaps elsewhere. The Welsh language too, despite William Hutton, remains lively and vigorous throughout Snowdonia.

Times have changed since the days when the people of

most lowland and semi-upland valleys were practically self-supporting, growing their own corn and grinding it into flour in the local mills. Most of the fields that grew the corn have long since reverted to unfenced sheep pasture. Bracken has invaded enormously. It is probable that most of the old fields were made by cutting down oak and birch woodland where bracken had been part of the undergrowth. It would persist for centuries round the edges of ploughed fields, and when the fields were turned over to pasture the bracken again spread across the ground. The trees have not come back, for their seedlings have little hope of surviving on a sheep-walk. But bracken thrives there, for though cattle would trample it and keep it in check, sheep do not. Nor do they eat it. Bracken likes acid ground and flourishes wonderfully on well-drained soils, sometimes covering miles of the slopes up to moderate heights, its underground stems spreading fast through deep, fertile soils. It is made little use of except as bedding for cattle. It provides cover for a few wild birds such as migrant warblers in autumn, but generally is not attractive to other birds because they find few insects among it. The preference that bracken has for the best soils is well known to the Welsh hill people, who express it in the saying:

Aur dan y rhedyn;
Arian dan yr eithyn;
Newyn dan y grug.
Gold under the bracken;
Silver under the gorse;
Famine under the heather.

Although compared with past centuries sheep are now superabundant, their right to the mountains and moorlands has not gone unchallenged. Since the 1920's more and more acres have disappeared under conifers and now we have vast forests that completely alter the character of their four main centres in the Park: the Dinas Mawddwy-Machynlleth area (Dovey Forest); the Mawddach and its tributaries north of Dolgellau (Coed-y-brenin); the Betws-y-coed region (Gwydir Forest); and the slopes from Moel

Hebog to Llyn Cwellyn (Beddgelert Forest). But first let us look back at the forests of past times.

The history of Welsh native woodlands is a fascinating one, beginning with the long-extinct woods of alder, birch and pine which grew along the coast in many places, a coast which was invaded by the sea in prehistoric times, say 3,000 B.C. The roots and stumps of these ancient trees are still preserved in the peat which lies under the foreshore. On the mountains also, especially on peaty moorland, are the copious remains of trees, especially birches, which flourished up to about 2,000 feet at a time when the climate was warmer and drier than it has been ever since. It is reckoned that the present cooler and wetter climate set in about 700 B.C. and that this deterioration stimulated the development of peat and waterlogged conditions. Trees could no longer survive on the high ground and they retreated to about 1,000 feet. Since about A.D. 200 a very slow improvement in the climate is believed to have been taking place; conditions are no longer as favourable to the growth of peat on the moors as they were and much of it has been eroded away. So it is possible that trees would by now have re-colonised parts of the upland region were it not that in the meantime man has stepped into the space the forests once filled, and by populating the hills with grazing animals has prevented any natural regeneration of trees. So it is only where walls and fences have been set up and trees deliberately planted that forests have reappeared in historic times.

If the climate wiped out the higher forests in prehistoric times, it was man who, throughout the historic period, decimated the thick forest cover of the valleys and lowlands. No doubt the most fertile valley bottoms have been cleared for farming for at least a thousand years. Since then the remaining forests have been gradually nibbled into and the former wide spread of deciduous forest is now reduced mainly to a pitiful remnant of slowly degenerating sessile oakwoods along some steep valley sides. The chief exceptions to this state of affairs are where enlightened landowners in the eighteenth and then the nineteenth centuries deliberately set out to restore their woodlands by a determined programme of fencing and planting. So by 1798 the Reverend John Evans could delight in splendid woods of broad-leaved and coniferous trees in the old forest of Gwydir near Betws-y-coed. There he saw flourishing groves of oak, ash, elm, sycamore and beech. He admired how 'the elegant spruce, the pensile birch and the rich scarlet berries of the mountain ash added a pleasing variety to the sylvan scene'. And he found the sweet chestnut so common and so natural-looking he could not believe authorities like Pennant who, quite correctly regarded it as an introduced species.

But though something was being done for forestry, there was still much neglect. In Caernarvonshire forty years after Evans we have Cathrall complaining: 'It is much to be lamented that planting is so generally neglected by the landed proprietors, as trees are not only useful but an ornament to every county, and particularly so in a cold, mountainous district, where they are so much wanted not only to hide the barren appearance of stone walls and naked rocks, but also for fuel and the various purposes of husbandry; at present there are some farms in this county, of 80 or 100 acres, without a single tree large enough for a fence-post.' He goes on to commend Lord Penrhyn and Lord Newborough for being praiseworthy exceptions to 'this culpable inattention to planting'. Other landowners who set out to beautify and enrich their estates with trees were mainly in the Vale of Ffestiniog; in the Dolgellau area; in the Dysynni valley; and around Bala.

The two world wars of the twentieth century, with their demands on home-produced timber, dealt harshly with such woodlands as existed. But since the 1920's we have experienced the revolution brought about by the programme of state forestry. Snowdonia, an area of wide, plantable spaces, has inevitably felt the full impact of the invasion. That this impact was not everywhere conducive to landscape beauty was soon evident when whole mountainsides received rectangular blocks of conifers quite out of keeping with the flowing lines of contours and hill shapes. People objected also that thickly packed conifers, making dark and dismal forests as contrasted with light-filtering woodlands of oak, beech and ash, could never be beautiful and in any case were quite alien to the traditional Welsh landscape. Years have passed since then and things have improved. It has long since become general practice to soften the edges of plantations with a border of hardwood trees; and now the idea of deliberate forest landscaping is firmly established. Conifers, however much one may dislike them, have to be accepted, for no other kinds of trees are so marketable on a big scale or would grow so quickly in the poor soils that cover most of upland Snowdonia.

The first question to be decided by the Forestry Commission in the twenties was which species should be planted, for when it came to afforesting the Welsh hills there was not a great deal of previous experience by which to be guided. So in many ways the work of the Commission has so far been a colossal experiment to find out which species are best for all the various habitats available in a hilly, Atlantic-facing region: exposed tops, sheltered slopes, frosty hollows, thin soils, deep soils, peaty soils and wet soils. Then it has been found that some species are prone to disease and others to damage by insects and voles. Of the many species tried, Sitka spruce is both the most successful and the most useful, and therefore the most commonly planted. It is a native of British Columbia, where it is one of the chief timbers used in the pulp industry. If allowed to, it grows for centuries and reaches great size. Unfortunately Sitka spruce is worse than any conifer for producing those black-looking forests which are so unpopular among defenders of landscape beauty. When it was realised that the cool, damp Welsh climate had more in common with that of western America than that of the continent of Europe, other American species besides Sitka spruce were increasingly used and now Douglas fir, western red cedar, lodgepole pine, western hemlock and various silver firs are all to be seen in the forests. Scots pine and European larch, formerly much planted, have proved disappointing. Scots pine has not stood up well to exposure; European larch is attacked by a fungal disease and has been replaced by a larch brought from Japan. Alas, this loveliest of conifers, Japanese larch, bright green in spring, warm red-brown all through winter, has in its turn gone out of favour because it has not fulfilled its earlier promise. In Snowdonia it grows rapidly to about forty feet but only very slowly from then on. Its use is now mainly restricted to shelter-belts, fire-breaks, and to giving a touch of colour around plantations of other species.

It is with mixed feelings that a naturalist must view these new forests. On the credit side they are a marvellous new habitat for some scrub and woodland birds, especially when the trees are very young. Birds attracted by the very first stages of moorland planting include meadow pipits and whinchats, and occasionally short-eared owls, for at that stage the grass is deep and voles may flourish exceedingly. Then the trees grow and bush out; the grass is smothered and these earliest colonisers give place to undergrowth-loving species such as willow- and garden-warblers, whitethroats, hedge sparrows, yellowhammers, robins, wrens and grasshopper warblers, birds which in pre-forestry days were very rare in the uplands. A few more years add several feet to the trees and now it is the turn of chaffinches, bullfinches, blackbirds and thrushes. Where the conifers are planted near heathery rocks, ring ouzels frequent them readily and sometimes nest in the trees instead of on their usual rocks. Two little finches are particularly favoured by conifers: redpolls are in all the forests, nesting mainly in trees from about eight to sixteen feet tall, especially Sitka spruce; and siskins have been breeding in much taller trees for some years in Gwydir Forest. Crossbills nest occasionally and may become regular as the forests mature. As the trees become tall and their lower branches are lopped they become useless to all the undergrowth-loving birds which originally colonised them and which go off to the shelter of newer plantations. Then it is mainly the turn of larger birds: jay, magpie, carrion crow, wood-pigeon, tawny owl and sparrow hawk; of small species only goldcrests, coal tits and a few chaffinches may be left as breeding species. Many mammals, too, are favoured by coniferous afforestation, among them pine-marten, polecat, fox, badger and red squirrel.

On the debit side of moorland afforestation there is the draining of bogs which may be the haunt of birds such as golden plover and dunlin, both of which are exceedingly thin on the ground in Snowdonia. Afforestation is known also to have destroyed or severely reduced the habitats of certain of the Park's rare plants such as the lesser twayblade, the bog orchid and the alpine enchanter's nightshade. But what many naturalists like least of all is to see a deciduous wood replaced by a coniferous wood and this is happening in many places these days. A conifer wood is always far poorer in birds, animals and plants than an oakwood or a mixed wood. When we lose an oakwood we lose a wealth of fauna and flora, as well as a type of woodland generally more beautiful than conifers; especially these days when larch is going out in favour of more and more Sitka spruce.

Forestry in Snowdonia is still a comparatively new and growing industry and is an undeniable social asset in a countryside where, apart from farming and quarrying, there is little prospect for young men to do anything but go to the towns for employment. Any industry in keeping with the rural tradition and which keeps country people in the country is welcome. Forestry does more: it provides healthy, outdoor work for many who would otherwise be sedentary or confined to the din and fumes of factories. Many forestry workers take a more-than-average interest in the natural world about them, those who work for the Forestry Commission being deliberately encouraged to do so. They learn not only about trees but they are sometimes well versed in the habits of the animals and birds of the forest, and can often provide most useful information about them. I, who have spent happy hours in the forests, watching and photographing birds, owe much to the kindness of forestry workers and officials and the interest they have taken in my activities.

With more and more planting continually going on by private enterprise as well as by the Forestry Commission, forestry is clearly destined to play an increasing part in Snowdonian life. At the same time a growing number of people are using Snowdonia for recreation. The Forestry

Commission has responded to this situation by the creation of two Forest Parks in Snowdonia, in the Beddgelert and Betws-y-coed areas; and by the provision of camping places, picnic sites and car parks, it welcomes all visitors who will treat the forests with respect. Furthermore, it has been agreed between the Park authorities and forestry interests that certain areas should not be planted.

CHAPTER THREE

Wild Plants

If you look up the mountain plants of Britain in the highly informative *Atlas of the British Flora*, you see that although their stronghold is naturally Scotland because the mountains there are higher as well as more northerly, yet there are many that reach as far south as Snowdonia. But there they end, most of them. The mountains of Caernarvonshire are the southern known British limit for alpine woodsia fern, alpine chickweed, arctic chickweed, alpine cinquefoil, arctic saxifrage, tufted saxifrage, chickweed willow-herb, three-flowered rush, hair sedge, jet sedge, alpine meadow-grass, and bluish mountain meadow-grass (*Poa glauca*). It is in Merioneth that six other species —alpine meadow-rue, moss campion, spignel, northern rock-cress, hermaphrodite crowberry and oblong woodsia— find their southern British limit, though the woodsia, recorded on Cader Idris, has not been seen for many years. A few others such as rose-root and purple saxifrage are found at one station still further south, in the Breconshire mountains, which reach almost to 3,000 feet; and still others such as mountain avens, mountain sorrel and alpine bistort grow in western Ireland at least as far south as Snowdonia; mountain avens grow in luxuriance there practically at sea level. It is curious that although certain typically upland species, notably mountain avens, grow on the west coasts of both Scotland and Ireland, they do not do so anywhere on the coast of Wales.

In contrast with all these northerners finding their southern limit in Snowdonia, we can offer only two southern species which have their northern limit here. They are the wavy St. John's wort, which grows in a bog near the Mawddach estuary, and the hairy greenweed, mainly a plant of the mild sea-slopes of Cornwall, 170 miles south, and really perplexing to find behaving like an arctic-alpine at nearly 2,000 feet on the crags of Cader Idris. No plant as far as I know is at its eastern British limit in Snowdonia and only one species is at its western. This is that curiously scattered, mainly westerly calcicole, the narrow-leaved bitter-cress, which grows on an ashwood-covered scree in Merioneth.

It is evident that Snowdonia's affinities are with Scotland and Ireland and that the division which geographers make between Highland Britain (the north and the west) and Lowland Britain (south, east and midland England) and which holds good for man, mammals, birds, fish and insects, extends also to the plants. Once you realise that Snowdonia's closest links are with north-west Britain you cease to be astonished at the scarcity even in lowland Snowdonia of a number of species that are common or fairly common not far away in the Welsh marches—species such as white campion, common St. John's wort, common mallow, musk mallow, creeping jenny, black horehound, white dead-nettle, butterbur and corn poppy. Perhaps the most striking example of all is the common cow parsley (*Anthriscus sylvestris*) which grows in such overwhelming abundance along the roadsides of England but which is comparatively rare along the roads of Snowdonia even in lowland valleys.

Then there are others which are distinctly happiest on base-rich soils and which, presumably for reasons of chemistry rather than geography, are rare on the predominantly acid soils of Snowdonia; plants like traveller's joy, field maple, guelder rose and quaking-grass. So we meet the very important three-fold division of plants into the calcicoles, or lime-lovers; the calcifuges, or lime-haters; and those which seem to be rather indifferent and are common on a wide variety of soils. The terms calcicole and calcifuge are very useful but it is as well to keep in mind that they are rather crude terms. There is no doubt that, lumped among the lime-lovers, are some—perhaps many—species which grow on the lime-rich crags not because there is lime but because there are other substances there which happen to be commonly associated with lime;

or because the rock on those crags is soft and friable and suits their requirements physically more than chemically. Also, we should remember that lime-loving and lime-hating may be parochial terms; they may apply very well in the British Isles yet not on the Continent. A species may appear as a strict calcicole or calcifuge in Britain but be far less choosey about its Continental habitats. Take the green spleenwort, for instance. This little alpine fern is reckoned to be exclusively a calcicole in Britain. Yet in the Alps it grows also on acid granitic rocks. This illustrates the ecological principle that when a species is on the outskirts of its range, as the green spleenwort is in Britain, it may be able to exist only in one very special sort of habitat, whereas nearer the centre of its range it may be far more adaptable. Other conditions, probably climatic, which are lacking at the perimeter of its range, favour it at the centre of its range and give it greater adaptability there. Though much has been written on this matter of what makes a plant a calcicole or a calcifuge, a great deal remains to be discovered.

Perhaps I should add a word on the special way in which the word calcicole tends to be used by those botanists who concentrate on the plants of predominantly acid regions such as Snowdonia. In chalk or limestone country botanists use the term calcicole for plants quite restricted to chalk or limestone, species such as the pasque flower, purple milk-vetch, dark mullein, yellow-wort and many others. But when you come on to the acid soils of Snowdonia you will hear local botanists apply the word calcicole to any species that cannot tolerate *very* acid conditions. I mention this loose but locally useful extension of the meaning of this word because it has been known to cause confusion among visiting botanists, whose eyebrows understandably go up when they hear described as calcicoles many plants which are hardly calcicoles in the full sense of the word: plants such as marsh hawksbeard, lesser clubmoss or flea sedge.

Much interest is added to the finding of wild flowers (or of anything else in nature) if we not only find and identify but also try to understand why a particular species lives where it does live and not elsewhere. A region like Snowdonia, whose rocks (see chapter 2) are remarkably varied and jumbled, lends itself particularly to this sort of approach compared with, say, some Continental mountain ranges that are entirely limestone. To get the best out of botanising in Snowdonia you should have or develop an eye for rocks as well. I do not mean you must become a geologist (though even a little geological knowledge and the possession of geological maps would help greatly), but that you must bear in mind that some types of rock are far richer in plants than others are and that you may be led to these better localities by knowing what to look for in the way of clues. It is well to know a few of the commoner lime-indicating species such as the ferns: green spleenwort, oak fern and bladder fern; or the mosses: *Tortella tortuosa*, *Neckera crispa* and *Ctenidium molluscum*; or the liverwort, *Preissia quadrata*; or flowering plants such as lesser meadow-rue and mossy saxifrage. No matter where you meet these plants you will be able to recognise where you are, ecologically speaking, even if geographically you are lost in a mountain mist! The rest of this chapter will cover a selection of the plants of the National Park habitat by habitat from the mountain tops to the coast.

The Summits

The most obvious feature of the summit flora is the fewness of both species and plants. In view of the terrible difficulties facing life on these exposed summits it is wonderful that anything lives at all. And even when a plant has braved everything that unkind soils and weather can inflict, it can still get beheaded by a sheep. All the same, plants exist right on the tops, some that merely struggle there but one or two which flourish. Usually the most luxurious plant is the woolly-haired moss which makes a thick, green-grey blanket among the rocks. Probably the decaying of this moss contributes more than any other plant to the thin humus that collects among the stones and allows the establishment there of the parsley fern, bilberry, cowberry and the other summit species. Of the three common clubmosses, the alpine and the fir reach right to the mountain tops, but despite its name the alpine is the less common of the two up there. The stags-horn clubmoss fades out several hundred feet lower. Quite a number of Caernarvonshire summits—Snowdon, Y Garn, Pen yr Oleu Wen, Carnedd Llywelyn, Carnedd Ddafydd and others have the least willow; in Merioneth it is on the two highest points—Aran Fawddwy and Cader Idris. Among the most likely summit grasses are common bent, viviparous fescue, annual meadow-grass and mat-grass. One member of the sedge family, the stiff sedge, belongs almost exclusively to the summits: in Merioneth, like the least willow, it grows on the top of Aran Fawddwy and Cader Idris; in Caernarvonshire it is probably on all the 3,000-foot summits and extends down some of the higher slopes as well. Many of the hard grey rocks, sometimes weathering very pale, are patched by delicate green lichens that indicate thoroughly acid conditions. A few of the ferns reach practically to the summits in crevices out of the wind, notably beech fern, hard fern, lady fern and the broad buckler; though none of these. it will be noticed, are what we usually think of as alpine ferns.

Mountain Grassland

Come down only a little from the weathered rocky summits and nearly everywhere you find yourself on the grassland that dips and rises over many miles of high country and eventually slopes away right down to the lowlands. Here and there it gives way to rock or heather-moor or patches of rushes, but generally this turfy cover forms the mountain skin and so contributes much to the beauty of upland Snowdonia. The seasonal changes so noticeable in the mountain scene are due to the colour changes of those plants which are dominant over areas large enough to be visible from far away. In spring broad belts of a lovely fresh green down the slopes show where bilberry is the prevailing vegetation. Dark-looking areas (in winter they look quite black) betray the heather-moor. Hillsides looking almost white in winter, but not from snow, are likely to be covered with either mat-grass or purple moor-grass, both of which turn very pale in autumn. Hills are commonly named from the striking colours their vegetation gives them, especially in winter. Wide patches of bracken, so richly red in the setting winter sun, have given to many hills the name Foel Goch, the Red Hill. The various black hills and black slopes (Foel Ddu and Llechwedd Du) may be named from their mantle of heather. White hills (Moelwyn), white patches of open ground (Llanerchigwynion), white fields (Dolwen) and so on, may well have got their names from a predominance of white grass. In some places natural changes or changes in land-use have made the old names no longer apt, but they survive. The fact that a Foel Goch gets planted with conifers and turns dark-green instead of red, does not alter its name. It will probably be called Foel Goch even if conifers cover it for evermore.

Though a beginner might be forgiven for supposing that in all this vast acreage of grassland we might expect to find nearly every kind of grass in the temperate world, in fact the number of species is extremely small. Of the hundred and fifty or so British grasses, four especially—sheep's fescue, common bent, mat-grass and purple moor-grass—are so perfectly adapted to life on acid mountain soils that they can close the door to most other species of grass, or keep them reduced to insignificant quantities. One or two other grasses such as wavy hair-grass and tufted hair-grass can be patchily abundant but very few others have managed to combat the oppressive blanket of the commonest grasses. But the heath rush does it very successfully; so does the moor sedge, growing tall and gracefully nodding where it escapes the sheep. All this vegetation is colourful in a limited range of fawns and greens and browns. But a little fugitive brightness comes from three plants, the yellow-flowered tormentil, the white-flowered heath bedstraw and the vari-coloured heath milkwort; all three are plants remarkably able to resist the competition of the grasses nearly to the summits. A very sweet fragrance comes off the heath bedstraw on some of the slopes, and on lower ground it is occasionally a food plant of the small elephant hawk-moth. When you get down towards a thousand feet, though the soils are still acid, there is less rain, more warmth and often better drainage, and the number of species quickly increases. Now there are eyebrights, violets, betony, sheepsbit, bitter-vetch, thyme, autumnal hawkbit, bell heather, mouse-ear hawkweed, broom, gorse, bracken and foxgloves. What a mercy it is that sheep do not eat every foxglove, for this is a magnificent species to find growing wild and sometimes abundant on the slopes.

As I have said in chapter 2, the rock lying under these mountain grasslands is less acid in some areas than in others and produces a sweeter bite for the sheep. Even if the sheep could not recognise a sweeter pasture if they saw one, any botanist could detect it for them by the presence there of plants such as yarrow, white clover or fairy flax. It is not until you get on to something approaching really calcareous ground that you see pastures and meadows coloured all over with a really rich variety of flowers. These are places to look for the frog orchid, the small white orchid, or the calcicole lady's mantle (*Alchemilla filicaulis*); and where you are likely to see common twayblades, greater butterfly orchids and that splendid adornment of the upland fields, the upright vetch, and with it sometimes the saw-wort. As it is of interest to note the absence of a species when that absence is surprising, I might point to two plants, very local in Snowdonia, which are quite frequent on similar-looking acid slopes in neighbouring Cardiganshire: the dyer's greenwood and the mountain pansy, two delightful wild flowers whose scarcity in the Park is as regrettable as it is mysterious.

Bogs

If the best upland pastures, from a shepherd's point of view, develop on the well drained, less acid soils, the worst are those growing on deep wet peat. Where rainfall is high and drainage impeded, peat can cover miles of the slopes, but it lies thickest on flat or hollow ground and especially in those wide shallow saucers that in post-glacial times were occupied by lakes long since choked by vegetation. You can recognise a peat bog from miles away by the blackness of bare peat, by the dark mantle of heather, by the beautiful white sheets of cotton-grass or, where erosion is in progress, by the way the peat is broken up into islands by countless winding channels. Peat bogs are usually avoided by the mountain walker, not only because they are wet and reputedly dangerous, but because they are so tiring to walk across. There is the prickly roughness of the heather, the incessant leaping across channels, and worst of all, there are those forests of tussocks that the single-headed cotton-grass forms when it has things all its own way in a peat bog.

A well developed, well preserved bog has a limited but very interesting flora. Where the peat is deepest, wettest and most acid we find little pools full of the bright green or sometimes reddish sphagnum or bog mosses—the dreaded 'bog-holes' of the moorland walker. In many such pools new peat is still being formed or, as botanists say, the bog is still active. Near these pools are often patches of almost bare wet peat so acid and inhospitable that not even cotton-grass, heather or crowberry can grow. It is in such open sites that we find the sundews, plants which need to extract little food from the ground for they can feed themselves by catching flies on their leaves. Only one species of sundew, the round-leaved, is common in Snowdonia. It very often grows on the top of the bog moss as well as on bare peat. Another, the long-leaved, is decidedly local, and the third, the great sundew, is very rare, being now known at only one site in Merioneth and not at all in Caernarvonshire.

But there are plants that are not insectivorous yet which flourish in these adverse conditions: white beak-sedge, which is rather local; bog asphodel, one of the loveliest of moorland flowers with its brilliant yellow spikes; cranberry, that creeps inconspicuously about the surface of the bog mosses; and the decorative cross-leaved heath. In bog pools not quite choked with sphagnum is often the insectivorous bladderwort which pokes up tiny yellow flowers just above the water surface; and the splendid bogbean, beautiful to see and so attractive to insects, including the marsh and pearl-bordered fritillaries. A very robust plant which sometimes almost smothers a peat bog and a great expanse of the surrounding moorland is the deer-grass, better called deer-sedge. Other robust species here are the bog myrtle, a delightfully aromatic shrub; and, unfortunately rare, especially in Caernarvonshire, the royal fern. Where bogs merge into slightly less acid but still wet ground there is an immediate increase in the number of flourishing plants; this is the habitat of attractive species such as:

Lesser spearwort	*Ranunculus flammula*
Marsh violet	*Viola palustris*
Marsh St. John's wort	*Hypericum elodes*
Petty whin	*Genista anglica* (not common)
Marsh cinquefoil	*Potentilla palustris*
Marsh pennywort	*Hydrocotyle vulgaris*
Marsh speedwell	*Veronica scutellata*
Red rattle	*Pedicularis palustris*
Lesser skullcap	*Scutellaria minor*
Devilsbit scabious	*Succisa pratensis*
Sneezewort	*Achillea ptarmica*
Heath spotted orchid	*Dactylorchis maculata*
Early marsh orchid	*Dactylorchis incarnata* (purple-red variety)
Lesser butterfly orchid	*Platanthera bifolia*

The outskirts of bogs are also the habitat of the aptly named tussock sedge (*Carex paniculata*), whose clumpy growth outdoes even the cotton-grass as an obstacle to moorland walkers. So it is fortunate that the tussock sedge does not normally spread itself over very large areas.

Mountain Crags

Most of these attractive bog-margin species flourish no higher than about a thousand feet. Above that the flora of the bogs is limited and most plant-seekers soon abandon them for the allurements of those base-rich crags that are the most famous plant-localities in the Park. Probably the wealth of plants on these cliffs comes from multiple causes. Obviously the bed-rock offers good mineral nutriment; the many ledges are a perfect arrangement for the collection of a rich store of humus; and down these cliffs so copiously supplied with rain there is constant and rapid drainage. Finally, and extremely important, not even sheep can get along vertical cliffs. It would be interesting to enclose an area at both top and bottom of a cliff to see whether, in the absence of sheep, the cliff plants would invade the grassland above and below.

I shall have more to say in the regional sections of the book about the flora of some of these upland cliffs. I think it may be most useful at this point to give a list of some of the flowering plants and ferns of two Snowdonian localities which, without being the habitat of rare species, have good samples of crag vegetation on them. One is the east-facing precipice of Craig Dulyn which rises to 2,200 feet at the east end of the Carneddau; the other is the gorge called Ceunant Mawr which is on the western approaches to Moel Siabod and rises to about 1,500 feet. Both these localities are on somewhat lime-rich rocks with plenty of water flowing or seeping down nearly always. These lists are not by any means complete but are intended as a guide to what might be found in about a couple of hours' searching. It will be seen that over three-quarters of the Craig Dulyn plants are also common lowland species. That they thrive on these high exposed ledges is a testimony to their adaptability, though many of them no doubt owe their ability to survive up there to the favourable soil they find on the base-rich rocks. Two of them, golden-rod and ox-eye daisy, like the rose-bay willow-herb, are represented on the high crags by special mountain forms. For instance, the short, compact, early-flowering golden-rod (*Solidago virgaurea* var. *cambrica*) particularly common on Cader Idris, seems very distinct from the taller, later-flowering one which also grows on the mountain ledges. But there are many intermediate forms.

Some Plants of Craig Dulyn. I have divided these species into two lists. The first list is of the typical mountain species, though it includes a few such as bladder fern, hard shield fern, beech fern and lady's mantle, which may occasionally be found nearly to sea-level.

Fir clubmoss	*Lycopodium selago*
Lesser clubmoss	*Selaginella selaginoides*
Parsley fern	*Cryptogramma crispa*
Green spleenwort	*Asplenium viride*
Bladder fern	*Cystopteris fragilis*
Hard shield fern	*Polystichum aculeatum*
Beech fern	*Thelypteris phegopteris*
Globe flower	*Trollius europaeus*
Lesser meadow-rue	*Thalictrum minus*
Alpine scurvy-grass	*Cochlearia alpina*
Common lady's mantle	*Alchemilla glabra*
Rose-root	*Sedum rosea*
Starry saxifrage	*Saxifraga stellaris*
Mossy saxifrage	*Saxifraga hypnoides*
Chickweed willow-herb	*Epilobium alsinifolium*
Ox-eye daisy	*Chrysanthemum leucanthemum*
Viviparous fescue	*Festuca vivipara*

The second list of Craig Dulyn plants is of the normally lowland species which nevertheless flourish on this and many other base-rich mountain crags. It will be seen that they far outnumber the alpine species.

Filmy fern	*Hymenophyllum wilsonii*
Hard fern	*Blechnum spicant*
Common spleenwort	*Asplenium trichomanes*
Lady fern	*Athyrium filix-femina*
Male fern	*Dryopteris filix-mas*
Broad buckler fern	*Dryopteris dilatata*
Polypody fern	*Polypodium vulgare*
Meadow buttercup	*Ranunculus acris*
Lady's smock	*Cardamine pratensis*
Wood bitter-cress	*Cardamine flexuosa*
Common violet	*Viola riviniana*

Heath milkwort	*Polygala serpyllifolia*
Slender St. John's wort	*Hypericum pulchrum*
Wood-sorrel	*Oxalis acetosella*
Holly	*Ilex aquifolium*
Western gorse	*Ulex gallii*
Broom	*Sarothamnus scoparius*
Bird's-foot trefoil	*Lotus corniculatus*
Meadowsweet	*Filipendula ulmaria*
Tormentil	*Potentilla erecta*
Wild strawberry	*Fragaria vesca*
English stonecrop	*Sedum anglicum*
Broad-leaved willow-herb	*Epilobium montanum*
Marsh willow-herb	*Epilobium palustre*
Wild angelica	*Angelica sylvestris*
Common sorrel	*Rumex acetosa*
Heather	*Calluna vulgaris*
Bilberry	*Vaccinium myrtillus*
Primrose	*Primula vulgaris*
Yellow pimpernel	*Lysimachia nemorum*
Foxglove	*Digitalis purpurea*
Heath speedwell	*Veronica officinalis*
Lousewort	*Pedicularis sylvatica*
Yellow rattle	*Rhinanthus crista-galli*
Eyebright	*Euphrasia officinalis*
Common butterwort	*Pinguicula vulgaris*
Wild thyme	*Thymus drucei*
Self-heal	*Prunella vulgaris*
Wood sage	*Teucrium scorodonia*
Harebell	*Campanula rotundifolia*
Heath bedstraw	*Galium saxatile*
Common valerian	*Valeriana officinalis*
Devilsbit scabious	*Succisa pratensis*
Golden-rod	*Solidago virgaurea*
Marsh thistle	*Cirsium palustre*
Hawkweed	*Hieracium* sp.
Marsh hawksbeard	*Crepis paludosa*
Common dandelion	*Taraxacum officinale*
Bulbous rush	*Juncus bulbosus*
Great woodrush	*Luzula sylvatica*
Heath woodrush	*Luzula multiflora*
Moor sedge	*Carex binervis*
Common yellow sedge	*Carex demissa*
Pill sedge	*Carex pilulifera*
Star sedge	*Carex echinata*
Flea sedge	*Carex pulicaris*

Purple moor-grass	*Molinia caerulea*
Wavy hair-grass	*Deschampsia flexuosa*

Some Plants of Ceunant Mawr. My Ceunant Mawr list would be over-lengthy to give in full so I include mainly those species which are typical of the base-rich rocks. Species such as common lady's mantle, meadowsweet, New Zealand willow-herb, starry saxifrage, butterwort and golden-rod, though quite happy on acid rocks, nevertheless often appear happier in base-rich localities. This is especially true of butterwort.

A calcicole moss	*Neckera crispa*
Common spleenwort	*Asplenium trichomanes*
Green spleenwort	*Asplenium viride*
Beech fern	*Thelypteris phegopteris*
Oak fern	*Gymnocarpium dryopteris*
Filmy fern	*Hymenophyllum wilsonii*
Globe flower	*Trollius europaeus*
Tutsan	*Hypericum androsaemum*
Meadowsweet	*Filipendula ulmaria*
Water avens	*Geum rivale*
Common lady's mantle	*Alchemilla glabra*
Orpine	*Sedum telephium*
Purple saxifrage	*Saxifraga oppositifolia*
Starry saxifrage	*Saxifraga stellaris*
Mossy saxifrage	*Saxifraga hypnoides*
New Zealand willow-herb	*Epilobium nerterioides*
Burnet saxifrage	*Pimpinella saxifraga*
Dog's mercury	*Mercurialis perennis*
Mountain sorrel	*Oxyria digyna*
Wych elm	*Ulmus glabra*
Primrose	*Primula vulgaris*
Butterwort	*Pinguicula vulgaris*
Lesser skullcap	*Scutellaria minor*
Common valerian	*Valeriana officinalis*
Golden-rod	*Solidago virgaurea*
Marsh hawksbeard	*Crepis paludosa*
Early purple orchid	*Orchis mascula*

In this list the plants that seem most demanding of the presence of lime are the moss, *Neckera crispa*; the ferns, green spleenwort and oak fern; and the flowering plants, purple saxifrage, mossy saxifrage, burnet saxifrage (an umbellifer), mountain sorrel and early purple orchid. Just as the Dulyn list can be taken as fairly typical of an average base-rich cliff flora in the Park, so the Ceunant Mawr list is typical of the flora of a rocky dingle cut through similar lime-rich rock. A list of the plants of Cwm Cowarch near Dinas Mawddwy which is about 30 miles south-south-east of Ceunant Mawr would be closely similar. Cowarch has in addition: bladder fern, hartstongue fern (rare), lesser meadow-rue, rose-root and Welsh poppy; but no one has yet recorded purple saxifrage there.

A striking feature of the alpine flora of the Park is the patchy distribution of certain species. A plant may be common in one cwm and quite absent from the next. To try to explain this we could say that many of these alpines are here petering out at the southern frontiers of their British range. Species such as:

Mountain avens	*Dryas octopetala*
Alpine chickweed	*Cerastium alpinum*
Twisted whitlow-grass	*Draba incana*
Northern bedstraw	*Galium boreale*
Arctic saxifrage	*Saxifraga nivalis*
Tufted saxifrage	*Saxifraga cespitosa*
Holly fern	*Polystichum lonchitis*
Alpine woodsia	*Woodsia alpina*
Oblong woodsia	*Woodsia ilvensis*
Hair sedge	*Carex capillaris*

These species may not be very adaptable and lack colonising vigour. Are they perhaps an increasingly doomed population as the last ice age recedes ever further into the past? But there is obviously more to the problem than this, for even some of the species that are still locally well established are peculiar in their distribution. Of these a good example is the northern rock-cress that is plentiful enough on Clogwyn Du'r Arddu yet only just about hangs on in Cwm Glas and is apparently absent from Cwm Idwal and Moel Hebog. Yet despite its absence from these likely places, it grows (admittedly very sparsely) on some rather unpromising rocks on Moelwyn. Or consider bladder fern and rose-root, two of the commonest Snowdonian alpines, yet they both seem to be absent from the rich community of Ceunant Mawr. Why? These are the perennial problems of the ecologists and will surely keep them busy a long time yet.

Mountain Springs

With all the rain that falls on them it is inevitable that Snowdonian mountainsides should bring forth a prodigious number of springs. Some of them make streams and gullies down the slopes. Others merely soak into the soil and make a patch of wet ground perhaps only a few square yards in extent. But small as they may be, these wet patches can be full of botanical interest, especially where the water issues from base-rich rock. Characteristic plants of these less acid mountain springs are lesser clubmoss, flea sedge, yellow sedge, separate-headed sedge and tawny sedge (local). And instead of the sphagnum mosses of very acid water, the mosses here are more likely to include such species as *Ctenidium molluscum* and *Cratoneuron commutatum*, as well as those sphagna, such as *Sphagnum contortum*, which are not extreme calcifuges. This sphagnum may be a typical associate of the very rare bog orchid (*Hammarbya paludosa*) which threatens to get ever rarer in Snowdonia through the destruction of its sites by drainage. On very high ground the number of species will be understandably fewer. But low down—and an ideal level is 800–1,000 feet—these wet patches, especially if they have long been enclosed as hay-fields, can be extremely full of variety. For instance, a corner of a meadow near Brithdir was found to have seven species of orchid as well as globe flowers, marsh hawksbeard, various sedges and a wealth of other meadow plants such as quaking-grass. The orchids were the heath and wood spotted, fragrant, dwarf purple, early marsh and both the butterfly orchids.

Lakes

I have never counted the lakes of the National Park but if we include the very small ones the number can hardly be less than a hundred and fifty. This is admittedly a naturalist's estimate, not an angler's, for a naturalist will be only too glad to count those completely weed-choked waters that an angler would scarcely flatter with the name of lakes, even though there are trout in some of them. Weedy lakes with their dragon-flies, water-beetles and birds are far more rewarding to naturalists than bare-margined lakes often are. All the same, even the clearest, stony-edged lakes can have a fascinating bottom flora. As five of these lake-bed plants have a similar growth-form—just a simple rosette of green, fleshy, finger-shaped leaves—they look very much alike as you look into the water at them. They are:

Shore-weed	*Littorella uniflora*
Awlwort	*Subularia aquatica*
Water lobelia	*Lobelia dortmanna*
Common quillwort	*Isoetes lacustris*
Small quillwort	*Isoetes echinospora*

When they are neither flowering nor fruiting it may be necessary to cut through a leaf and see it in cross-section in order to distinguish the species. Only the lobelia has obvious, above-water flowers. The flowers of shore-weed and awlwort are tiny and usually formed under water. Quillworts are spore-producers and therefore flowerless. Of these five the shore-weed is most abundant, often making a wide green carpet under the water of shallow lake margins. But lobelia is also frequent and may completely occupy the bed of a shallow, small lake. The awlwort, minute member of the cabbage tribe, is something of a rarity and is probably decreasing. Of the two quillworts, one (*Isoetes lacustris*) is quite frequent; the other (*Isoetes echinospora*) is found in few of the lakes. In some lakes you can see a decided zonation of these species according to water depth, just as sea-weeds are zoned: nearest the edge is often a sward of shore-weed, then a belt of lobelia leading to the quillwort zone in the deeper water. These lake-bed plants can be a delight to look down on from the crags on a still, bright day, for with the flotegrass (*Glyceria fluitans*), the floating bur-reed (*Sparganium angustifolium*), and the various starworts and pondweeds, they can turn the whole lake into colourful patterns of brown and light green. Although these plants can tolerate the acid water of the mountain lakes, they are usually more abundant in those lakes that receive the drainage of the base-rich rocks; which accounts for the very fine lake-floor carpet of plants in Llyn Idwal and the two similar lakes, Aran and Gafr, under Cader Idris. Purity of water could well be another factor in the survival of a struggling and sensitive species like the awlwort. Only slight pollution would perhaps tip the balance against it.

Most popular and showiest of all lake plants are the water-lilies—popular, that is, except among anglers who much dislike the powerful and embracing stems and great leaves with which these lilies invade large areas of some lakes. The water-lilies of Snowdonia are the same two species common in the English lowlands, the white water-lily with its very large flowers, and the yellow with its smaller flowers but larger leaves. Both are fairly frequent on lakes up to about 1,500 feet, and sometimes grow side by side. A yellow water-lily in Talyllyn Lake is a hybrid between the common yellow and the least yellow water-lilies, and is of particular interest because neither of its parents exists in the lake though presumably both were there formerly. It is intriguing to speculate how long ago this hybrid arose there because the one parent, the least yellow water-lily (perhaps formerly more widespread?) is a rare plant in Britain, being confined to central and north Scotland, apart from one or two waters round Ellesmere in Shropshire.

I should add a special word about Bala Lake as it is so much the largest sheet of natural water in the Park. It is a long, windy and often rough lake and so has not developed reed-beds or other above-water vegetation except in a few sheltered inlets. These are among the very few sites in Snowdonia for the sedges: bladder sedge (*Carex vesicaria*), water sedge (*Carex aquatilis*) and *Carex acuta* (variously called in English the graceful, the tufted, the slender-spiked or acute sedge). But, sad to record, the lowering of the lake-level a few years ago has almost drained these inlets and the sedges seem doomed to disappear. Shore-weed grows in some quantity at the north-east corner of the lake.

The Wild Plants of Civilisation

Coming down off the hills to the lowlands we pass into the deciduous woodland zone with its special flora. But as I shall be dealing with woodlands in connection with several of the nature reserves I leave this subject for chapter 6 and come straight to the valleys that characteristically lie under the high woods, valleys where lonely farms and hamlets begin and where there are roads that lead down to other, wider valleys, and so eventually to larger villages, towns, main roads and a few railways. I wish I could say canals too, but there are no canals in Snowdonia: a pity, because a disused canal winding through an unpolluted countryside can be a valuable botanical asset as well as being rich in animal life.

The vicinity of man is the realm of aliens and garden escapes. Some are the relics of former herbal uses and insist on going on being man's companions; others are modern garden species that have been thrown out or have left of their own choosing. I am thinking of plants like spindle, comfrey, Welsh poppy, burdock, evergreen alkanet, slender speedwell, soapwort, wormwood, greater celandine, fennel, horseradish, gout-weed, Good King Henry and Japanese knotweed: all plants that tell you that you are approaching a farm or a hamlet and which rarely get far from civilisation. But a blue mat of slender speedwell astonished me once on a bank miles up the Dovey valley above Dinas Mawddwy, well away from any house. If it can get as near to the wilds as that, where is this spreading species ultimately going to get to? The Welsh poppy is unique as being the only native Welsh alpine—it belongs properly to base-rich mountain ledges—to have become a village weed. Several other poppies are weeds of railway banks, where they find themselves in company with a crowd of other casuals like Pyrenean cranesbill, small toadflax, pale toadflax, early yellow cress, apetalous pearlwort, rosebay willow-herb, evening primrose and lamb's lettuce. And among this jostling, elbowing crowd of cosmopolitans a few native plants also get in here and there, especially by the sea: from the dunes have come yellow sheets of biting stonecrop, and from the shore many shy but persistent little clumps of Danish scurvy-grass. Wherever man throws out rubbish or leaves a patch of waste ground for nature to play with, you can find another bunch of tough invaders. Some of those that come successfully through the rough-and-tumble of rubbish-tip or wasteland competition are railway successes too. In this list we can put:

Viper's bugloss	*Echium vulgare*
Vervain	*Verbena officinalis*
Hemlock	*Conium maculatum*
Small nettle	*Urtica urens*
Dyer's rocket	*Reseda luteola*
Canary grass	*Phalaris canariensis*
Narrow-leaved everlasting pea	*Lathyrus sylvestris*
Mugwort	*Artemisia vulgaris*
Common mullein	*Verbascum thapsus*
Black horehound	*Ballota nigra*
Lesser swine-cress	*Coronopus didymus*

There is also that American willow-herb (*Epilobium adenocaulon*) that is spreading so rapidly everywhere. Even two orchids come close to this category (but without being in the least aggressive): the broad-leaved helleborine, which in Snowdonia is usually at roadsides; and the rare sword-leaved helleborine which, at its one locality in the Park, spills over from a roadside down on to a railway bank.

Estuary and Shore

The Park is well provided with river mouths. There is the superb estuary of the Mawddach, the much smaller but very interesting Artro; and there are parts of the Dysynni, the Dwyryd and the Dovey within the Park. In the north the estuary of the Conway goes alongside the Park boundary for several miles. Though they all have one thing in common, namely the periodic invasion of salt water, each estuary has its peculiarities, and the botanist who has also an interest in the ever-changing shapes of the saltings in their relation to the vegetation has plenty to look at and discover about all these North Wales estuaries.

As for individual species, one must in these days pick out the rice-grass (*Spartina townsendii*) as of outstanding importance. This infamous spreader (which helped to reclaim the Zuider Zee) was first planted in the Dovey estuary in 1920 in the hope that it would establish itself and turn open mud-flats into grazeable sward. It succeeded beyond, I imagine, the most optimistic hopes of the land-owner who planted it, and today it extends in a broad belt for several miles down the estuary. It has also gone up the coast and, mainly in the 1950's, established itself in great beds in other estuaries, especially the Mawddach. It grows tall and even and by August looks very much like a prosperous crop of wheat, the similarity being heightened in autumn when it turns golden-brown and is very lovely in the evening sun. Though a hybrid spreading mainly by underground runners, it is a fertile hybrid and in some years produces abundant seed that is attractive to birds, notably reed buntings, in late autumn. Its thick growth is also very welcome cover to many ducks and waders. All the same, it is, on balance, a most undesirable plant from a naturalist's point of view: it not only ramps over other weaker plants but also forms a thick mat over open mud and so destroys the feeding grounds of the estuary birds. So devotees of salt-marshes look askance at the all-invading rice-grass and wonder how long it will be before all the glasswort, seablite, spurreys and estuarine sea-weeds are lost in the jungle.

Botanists who are used to the east coast of England and therefore take for granted large quantities of sea lavender and sea wormwood will be surprised to find them rare or absent on the salt-marshes of Snowdonia. But the other usual estuary species are common enough: the swards of sea meadow-grass (*Puccinellia maritima*) and red fescue (*Festuca rubra*) and also thrift, sea plantain, sea arrow-grass, common scurvy-grass and parsley water dropwort; the typical salt-marsh rushes and sedges; and in some of the brackish pools the forests of slender green fronds of the tassel pondweed where shoals of darting little fish seek shelter when you approach. The heads of some of the estuaries, notably Mawddach and Conway, zone off into extensive reed-beds and in and around the ditches the celery-leaved buttercup and other plants of brackish conditions are common. A water crowfoot, *Ranunculus baudotii*, is found, but rarely, in some brackish ditches.

I will leave to the chapter on nature reserves most of what I have to say of the shore and dunes. Suffice it to say here that cliff, sand, shingle and dune plants are well represented along the Merioneth coast, and to mention a few plants that are not in the duneland nature reserves. There is, for instance, a scattering of samphire (*Crithmum maritimum*) on the rocks. But that other umbellifer, alexanders, is not nearly so common on the Merioneth coast as it is elsewhere round Cardigan Bay or on the north coast of Wales. Vernal squill is also curiously rare. The wild madder (*Rubia peregrina*) which in Snowdonia is very close to the northern edge of its British range, is another scarce species. Lanceolate spleenwort grows mainly around Barmouth—some of the old fern-collectors knew it as 'Barmouth spleenwort'—and there it often seems to be associated with manganese-bearing rocks. Finally there is the sea spleenwort, the only Snowdonia fern that prefers the salt sea winds to any other habitat: but it only just gets into the list on the strength of a very few plants in the cliffs south of the Mawddach.

In a region so large and, in places, so wild as Snowdonia there are obviously many botanical discoveries that can still be made by the enthusiastic amateur. I do not mean only the finding of rare plants, though some undoubtedly remain to be found. What I have in mind particularly is the sort of work that Price Evans did on the mountains of Merioneth in relating communities of plants to the lime-rich and acid rocks. He used to follow these plant communities in a most fascinating way for miles across country with a sort of bifocal, geological-botanical vision. His papers in the *Journal of Ecology* are a lead to what a keen amateur can still do on similar ground throughout Snowdonia. Price Evans included whole vegetations—lichens, mosses, as well as flowering plants and trees—within his field and devoted the leisure time of many years to his work. But many shorter and very useful studies could be made of the ecology of individual species; and it hardly matters which you choose for there is something new to be discovered about them all. Why a plant grows here and not there is going to puzzle us a long while yet.

As for seeking rare plants, obviously in such a long-searched region as Snowdonia the chances of finding anything fresh are limited. Two types of discoveries of rare plants are possible: first, the rediscovery of 'lost' plants, recorded perhaps a long time ago and not seen since; second, the discovery of species entirely new for the Snowdonia list. The best hope is in the rediscovery of the 'lost' species. This is a sport for those who can most easily believe in old records. And certainly some of them have been too hastily rejected. For example, the hairy greenweed, first recorded from Cader Idris in 1800: it was rejected in 1835 as 'an error' only to be rediscovered in 1901. Other 'lost' Merioneth alpines re-found in the last few years are juniper, alpine saw-wort, moss-campion, alpine meadow-rue, northern bedstraw, least willow and stiff sedge. Still others found, or claimed as found in

Merioneth but not now known, are holly fern, oblong woodsia fern, pillwort, marsh clubmoss, grass of Parnassus, pale heath-violet, flax-leaved St. John's wort, alpine meadow-grass and quite a few other less likely species. Valuable guidance on these matters can be found in two excellent sources: P. W. Carter's paper on *Botanical Exploration in Merionethshire*; and the *Contribution to a Flora of Merioneth* by Peter Benoit and Mary Richards. A similar account could be given for Caernarvonshire. Though Evan Roberts, the Nature Conservancy's warden, has worked hard for years on the alpine plants, he would be the last to suggest that he has found everything. For instance, the long-recorded but now unknown Killarney fern still challenges him. Another old record, that of the marsh clubmoss, tended to be discredited until the plant was rediscovered a few years ago. The tufted saxifrage was believed by many to be extinct in Snowdonia but has lately been found again in exceedingly small quantity, its few seedlings being devoured by the mountain slugs. Alpine bistort, found by Lhuyd in the seventeenth century, went unrecorded until Mary Richards and Evan Roberts found it this century. The possibility that *Saxifraga rosacea*, a rare species of mossy saxifrage, also reckoned as extinct in Snowdonia, may yet be found again, is believed in by at least one optimist. The fact is that when a botanist's prejudices are against all hope of finding a plant, he is very unlikely to find it even if he looks. Its discovery may then be made by pure chance by a butterfly-chaser or bird-watcher! So perhaps one day it will fall to a bird-watcher to reinstate the chickweed wintergreen (*Trientalis europaea*) as a Welsh plant. That really would be recovering the past, for as far as I know this species has not been claimed for Wales since it was first reported in Parkinson's *Theatrum Botanicum* of 1640: 'Also in the Beeche wood in Scotland, as it is recorded by Bauhinus, who saith Dr. Craige sent it him from thence, and on the mountaines in Wales likewise.'

CHAPTER FOUR

Mammals, Reptiles, Fish and Insects

Mammals

What is most distinctive about the mammal and bird populations of Snowdonia is the survival here of several large predatory species, some in considerable numbers, which have been exterminated or nearly so in most lowland districts of Britain. Of course we have lost some, and very long ago. We lost the bear probably in the eighth century, the wolf in the sixteenth, but the wild cat not until the nineteenth. No golden eagle has nested here for perhaps 200 years and no doubt kites and harriers used to be far less rare than they are now. But we still have six different kinds of birds of prey breeding and two others probably do so occasionally. And we have seven species of carnivorous land mammals, all common except one. The survival of all these predators is due to one all-important factor: the continued existence, natural in a predominantly mountain region, of a great many wild or semi-wild habitats into which the predators, though they have been sorely persecuted in the valleys, have always been able to retreat. It is not very difficult for a ruthless gamekeeper to wipe out the larger predatory mammals of a lowland wood in agricultural country. But it is a different matter if the wood is on a steep mountainside scattered with rocks and small crags, and above the wood there is a wilderness of block scree full of huge caves and fissures, and above the scree is a line of cliffs, and beyond the cliffs stretch the open moorlands. In that sort of habitat, which is common throughout upland Snowdonia, the odds are clearly in favour of the survival of at least a few of the predators.

That the wild cat did not manage to survive in some remote mountain region—in the wild, heathery rocks of the Rhinog, for instance—is not easy to explain. Presumably its extermination reflects the special hatred that farmers and shepherds had for it, so that they went to extreme lengths to get rid of it. Settled rural communities have always protected their flocks and poultry against predators. The churchwardens' accounts of two or three centuries ago are a valuable source of information about the history of our ravens, falcons, martens and foxes because of the frequent payments recorded in them for the destruction of these flesh-eaters. In the accounts of 1822 of Talyllyn church in the Dysynni valley we read: 'Humphrey Jones for killing three young ravens 2/–,' and in 1837 we find: 'Paid Ellis Williams and R. Owen for killing Bella (marten) 2/6.' All through the nineteenth century this persecution went on and with increased intensity because more and more gamekeepers were now flung into the battle. By the beginning of this century most of the larger raptors, bird or mammal, were beginning to get scarce even in the wildest Snowdonian uplands. It was the First World War which, by diverting most gamekeepers to other occupations, pretty certainly saved several predatory creatures from extinction by removing persecution just in time. After that war gamekeeping was in many areas never resumed on the pre-war scale and since then there has not only been less persecution because less gamekeeping; there has been a pronounced shift in public opinion in favour of conserving wild life, even the species traditionally classed as vermin. So it is greatly to be regretted that the wild cat, which apparently survived in Snowdonia until so very recently, could not have held on for those further few decades that would have brought it into what we hope are more enlightened times. A few wild cats would have been a great enhancement of Snowdon Nature Reserve.

Because of its rarity and its beauty the pine-marten is surely Snowdonia's most distinguished mammal. But, alas, in many years of wandering and searching in the wildest places I have not yet had the luck to see one. Very few people ever do. This is hardly surprising when one remembers that, as well as being extremely scarce and mainly nocturnal, the marten is apparently a creature of no fixed abode most of the year. As appears from the few very widely scattered records, though it may turn up almost anywhere, it is useless to watch for it at any particular place unless you happen to know where its nest is. In fairly recent years the places where martens have been seen in Snowdonia include the country near Beddgelert, Betws-y-coed, Bontddu and probably Arthog. I know of no recent records from Aran Fawddwy but there seems no reason to suppose it should not have survived in so mountainous an area. Plantations of the Forestry Commission (where martens have been seen and even accidentally trapped on very rare occasions) and rocky upland districts seem to be the likeliest habitats now for this elusive rarity. It is difficult to say precisely how long the marten has been so rare. Pennant knew it as common in the eighteenth century when its valuable fur was 'much used for linings to the gowns of magistrates'. We can fairly assume that martens grew steadily scarcer during the nineteenth century, though even at the end of it there were enough of them at least on the south Merioneth hills to be still a deliberate object of the chase.

I have seen what I believe were marten tracks in the snow near Machynlleth, and Bolam describes how he followed such tracks from the middle of the village of Llanuwchllyn to the top of Aran, noting that the marten had killed and eaten a rat on the way. This foray of at least four miles, involving a climb of over 2,000 feet, is an illustration of the marten's roving nature. Other much longer trails through the snow are on record. Bolam actually saw a marten on three occasions not many miles from Llanuwchllyn, his best view being when his attention was attracted to one being mobbed by rooks that had come up to feed on Aran Fawddwy. Another excellent view of one was had by Dr. J. H. Salter on the top of Rhobell. He, like Bolam, was botanising at the time and again it was excited birds, this time wheatears, that drew attention to the marten. It would seem advantageous to would-be marten-spotters to be both botanists and bird-watchers! Salter's marten was evidently a particularly colourful one, for it appeared to him 'as red as a squirrel'. Most marten-furs are a very much darker brown than that, some being almost black. Pennant had no doubt about the marten's pride of place: 'This is the most beautiful of the British beasts of prey: its head is small and elegantly formed: its eyes lively: and all its motions show great grace as well as agility: when taken young it is easily tamed, is extremely playful and in constant good humour.'

Snowdonia's other really distinctive mammal is that other beautifully furred member of the weasel family, the polecat, which is always darker than any marten and in fact, except for white facial markings, looks at a distance completely black. This is because the longest hairs of its fur are quite black, a sleek and glossy black; but when these black hairs are turned back you see that the thick under-fur is cream-coloured. This pale colour is very noticeable when a polecat twists its long sinuous body in turning round and so parts the rather sparse black hairs. Striking colour-varieties—usually called red polecats—vary from straw-coloured, sandy and ginger to a bright fox-red. This variety was first recorded south of the Dovey early this century. It reached Snowdonia in 1927, four being killed in traps in the Aberdovey-Towyn district. Red polecats (among large numbers of normal ones) continued to be trapped during the 1930's but then seem to have died out there. But as this variety appeared once, there is no reason why it should not do so again and, as has frequently happened in the past, these red polecats could again be mistaken for martens.

Widely distributed in Wales but not found elsewhere in Britain except that it seems to be now spreading into the Welsh marches, the polecat is common in Snowdonia and especially so in Merioneth, having apparently increased this century due to a lessening of persecution. It is a little smaller than the marten but much bigger than the stoat. It lives in a variety of country from coastal sand-dunes and lowland farms up to the highest conifer planta-

PLATE I *Above*, man's early past is copiously signposted in Snowdonia by remains particularly of the Late Stone, Bronze and Iron Ages. Here is the great capstone of a Late Stone Age cromlech built into one of the long cairns of Carneddau Hengwm near Barmouth. The hill in the background is Moelfre *Below left*, fossil shell (a brachiopod of the genus *Dinorthis*) from the summit of Snowdon, a reminder that in the Ordovician period when its rocks were formed of alternating sedimentary and volcanic strata, Snowdonia was part of an ocean bed

PLATE II *Above*, some of the signs of glaciation in the last Ice Age are erratic boulders left by the melting ice such as this one near Helyg and ice-scratched rocks (*below*) in Nant Peris

PLATE III **Two so-called Roman remains.** *Left*, The Roman Steps are in fact part of a medieval road from Harlech to Bala. They begin near Llyn Cwm Bychan and climb up a narrow defile on the north side of Rhinog Fawr

Below, The Roman Bridge, or Pont y Benglog, which spans the falls on the River Ogwen at the west end of Llyn Ogwen is an eighteenth-century pack-horse bridge. It has been superseded by the bridge above it, which carries the A5

PLATE IV *Left*, winter sunrise on Snowdon from the Mymbyr lakes. From left to right the peaks are Lliwedd, Snowdon, Crib Goch and Crib y Ddysgl. *Below*, the east face of Snowdon from Llyn Llydaw. By the lake are the ruins of nineteenth-century copper mines

PLATE V **Mountain Flowers.** *Above left*, Snowdon lily (*Lloydia serotina*) is a rare plant of high, base-rich cliff ledges of Snowdon and nearby mountains; *right*, detail of the flower
Below left, on the Glyder and Carneddau ranges mountain avens (*Dryas octopetala*) reaches its southern British limit
Right, Welsh poppy is a widespread mountain ledge plant of Snowdonia

PLATE VI *Above left*, Rock stonecrop (*Sedum forsteranum*). *Right*, Least willow (*Salix herbacea*) a creeping shrub of high altitudes
Below left, Awl-wort (*Subularia aquatica*) a small water plant of a few Snowdonian lakes. *Right*, Marsh helleborine has only two stations in Snowdonia, the sand dunes of Morfa Dyffryn and Morfa Harlech

PLATE VII *Above*, bog rosemary (*Andromeda polifolia*), a pink-flowered member of the heather family rare in most of Britain and in the Snowdonia National Park found only in southern districts.
Below, vernal sandwort, a common species of base-rich rocks on Snowdon and some nearby heights. In Merioneth it is very rare

PLATE VIII **Paths to Snowdon's Summit.** *Above*, looking north-east from the top towards Pen-y-pass, showing the Pig Track climbing along the slopes of Crib Goch, the peak in the centre. The Horseshoe Track follows the knife-edge along the top of the ridge. *Below*, the view south-east from the summit showing the Horseshoe Track continuing round the sharp edge of Lliwedd with the Watkin Path coming up (*right*) from Cwm y Llan

tions. It habitually hunts near villages and farms and for this reason is rather a frequent road victim. Indeed, the number killed by traffic is our best indication of how numerous polecats must be. In 1935 an inquiry by E. H. T. Bible revealed that 90 polecats were killed that year by gamekeepers in the Dovey-Dysynni area. Like martens, polecats used to be hunted with hounds last century, but not all hounds, it seems, would hunt the polecat. N. W. Apperley described how he failed to get otter-hounds to hunt polecats but was more successful with a specially trained pack of old southern hounds, a breed renowned for keenness of scent. In quiet places the polecat, like many other nocturnal animals, is occasionally to be seen out in broad daylight. I have seen one nonchalantly crossing my garden, hunting right and left as if for voles, and passing carelessly within three yards of me. Another I saw was obviously hunting rabbits, going systematically from one burrow to another. Sometimes at night our dog has met and chased polecats for about fifty yards and has returned so saturated with the stench which, when excited, they emit from their glands that we have been scarcely able to bear his presence indoors for a couple of days. A correspondent who lives near Harlech has recently described to me how one morning (July 26, 10.30 a.m.) she saw four or five young polecats at the foot of a bank in a wild brambly patch at the end of her garden. They were frolicking and tumbling about and biting one another's ears, and showed no fear of her presence nor of that of her friend whom she called to look. A hole in the bank was obviously the polecats' home, for all the vegetation round it had been trampled down. Meat and bones were put out every night and the adult polecats took them regularly and once in good light at 9 p.m.

Neither stoats nor weasels seem to be as common as they were a few decades ago and it seems likely that both are at present outnumbered by polecats in many parts of the Park. But it is easy to underestimate the numbers of a shy, mainly nocturnal species. From the number of reports it is evident that quite a proportion of stoats turn white in winter in Snowdonia.

In rocky places both stoat and weasel range fairly high into the mountains. Not so that large member of the family, the badger. Its main strongholds have always been the deciduous woods of the lowlands, where the soils are deep and yielding enough to permit its great burrowings and earth-removals. In upland woods, often sheep-grazed, rocky and lacking undergrowth, badgers are much scarcer and tend to find what cover they can in the banks of dingles. Outside the hill woods you occasionally come on sets on open hillsides but not, in my experience, above the bracken line. Only in the forestry plantations are there appreciable numbers of badgers in the hill country, for among the conifers they find cover, food and workable earth. Science classes the badger as a weasel; but in the Celtic world the badger has always been thought of as a wild pig—*mochyn ddaear*, which means 'hog of the earth'. In the Celtic world in early medieval times badgers were hunted for their flesh and their hind quarters made into hams, a dish that was served on ceremonial occasions as a mark of esteem for distinguished guests.

Those two victims of present-day hunting, otter and fox, are widespread. But otters are now scarce. They are mainly on the deeper lowland rivers, estuaries and coasts where fish are most readily found, but they frequently go up rocky streams and gorges to visit upland lakes. No animals are more furtive and shadowy in their movements than otters, and though quite numerous, they are seldom seen. For most of us only their whistling at night and their pad-marks in soft ground tell of their passage up and down the streams. Like all the other predators, the otter likes to come out in daylight when it is safe to do so. One day in the middle of the wide sands of the Dovey estuary at low tide an otter boldly ate a flat-fish although I stood looking at it from only fifty yards away.

Foxes, unlike otters and badgers, haunt the highest places of the mountains, and of all the nocturnal mammals are the most frequently seen by day. I would say that mountain foxes have three favourite types of territory: block scree, peat hags and cliff ledges. If you surprise a fox in block scree you have little hope of a good view;

you may see him leap lightly from one boulder to another and then he is gone among the rocks. Foxes love to sleep out in the sun and in peat hags they curl up in the heather in a warm spot out of the wind; in such a place you usually flush your fox unexpectedly; you see him start up with alarm and run away over the moor, quickly yet without appearing to be going full out, and pausing once, perhaps twice, to look back at you before he goes over the horizon. I have several times, in looking down heathery cliffs for peregrines' and buzzards' nests, seen a fox lying on a ledge below me. One of them was soundly asleep close to the roar of a waterfall, which was an imprudent spot to choose because I was able to slip softly down through the heather to within a few feet above him and drop a little pebble on him to wake him up! A mean trick perhaps, but he was a beautiful sight leaping away down the terraces with his fine tail streaming behind him. There is an old belief that two distinct races of fox live in Wales: the mountain fox, large, strong, swift and grey; and the lowland fox, smaller, redder and not such a good runner. The grey mountain fox used to be called *milgi*—the grey-hound; the red lowland fox was *corgi*—the cur-fox. Modern science does not, as far as I know, accept this distinction but there may well be something in it. Certainly Welsh mountain foxes have been taken in numbers to be released in the shires of England in the belief that they give the hounds a better run than lowland foxes do. A great deal of trapping, shooting and cub-destruction of foxes goes on incessantly in Wales and very large numbers are killed annually. Despite this, the fox is still plentifully with us, so presumably it can stand such drastic thinning. Controversy rages and always will rage about how much damage foxes do in the lambing season. The poultry-keeper's case against the fox is much clearer. Not that foxes are entirely flesh-eating. Far from it. Their diet is varied and, as you can easily see from their purple droppings in summer, they eat quantities of bilberries. One final point: in severe winters there are always stories of 'packs of foxes' coming down from the hills to raid lowland farms. But we should remember that the coldest weather usually coincides with the fox's mating season when, regardless of the weather, it is not abnormal to see quite a party of dog-foxes squabbling over a vixen. These gatherings could easily be exaggerated into something pretty hair-raising by some imaginative raconteur who has had a couple of pints in the village inn.

Just as foxes eat fruit, so may all the flesh-eating mammals and birds occasionally vary their diet. But taking the year as a whole, many of them depend on the availability in large numbers of the smallest mammals, what we might call the three-inch group, for that is about the average length of the multitudinous mice, voles and shrews that live in the woods and grasslands throughout the valleys and mountains. But these small rodents and insectivores are even more difficult to observe than the carnivores, for they are not only largely nocturnal but even when they move about by day they are nearly always hidden in vegetation. You may hear a shrill squeaking, especially from the shrews, but that is all. The result is that in a varied and mountainous area such as Snowdonia there is still much to be learnt about the numbers, distribution and altitudinal range of these small species. So there is a whole field of original work awaiting some investigator prepared to do a bit of exploring and mountaineering. For the casual walker, especially if accompanied by a dog, the shrews are the easiest species to detect because of their habit of dying above ground. Nothing decays more quickly nor more odorously than a dead shrew and a dog will often lead you to one by smelling it out and rolling on it. By this means you soon learn that both the pygmy and common shrews are to be found to the tops of the mountains. But a great deal remains to be found out about the third species, the water shrew. Beyond the obvious fact that it is far rarer and far more local than the other two, very little is really known about its distribution in Snowdonia. The few records H. E. Forrest obtained were all from lowland districts or near the edges of the Park: Conway valley, Bala, Dovey and Penrhyndeudraeth. Now, as then, records are still wanted for all inland districts, especially upland ones. The only high-level water shrew

I have met with was a very beautiful creature, jet black above and pure white below, which I watched running rapidly about the margin of a moorland lake at about 1,500 feet searching for food between the stones. In the lowlands, water shrews are known to live in woodlands far from water. It would be interesting to discover whether in mountain districts they can likewise be independent of lakes and streams. Bolam, though he only ever saw one water shrew alive in the Llanuwchllyn area, found their bones in the pellets of owls, hawks and ravens. This is probably the best way to detect the presence of so elusive an animal. As for that other little mammal of the water-side, the water vole, I have found it scarce in most of Snowdonia compared with the numbers one sees for instance in lowland districts of east Wales and the marches. Earlier this century it seems to have been commoner. The highest I have heard of is the one Forrest recorded at 800 feet.

There is no doubt that an important link in the chain of life on high ground is the enormously abundant short-tailed vole (*Microtus agrestis*), for it is an important food of kestrels and moorland owls (all four native British owls hunt at times on moors) and probably of mountain foxes, martens, stoats and weasels. Not much seems to be recorded about its relative numbers at different heights but presumably it exists in small numbers right to the mountain tops. It is commonest on the hills where the grass, heather or bilberry is thickest. If a stretch of mountainside is enclosed to exclude grazing animals, voles soon increase in the deeper grass that develops there. Voles also thrive in bogs overgrown with purple moor-grass, a deciduous grass that does not at first sight seem to offer much in the way of winter sustenance. But in the heart of its thick, dead-looking, whitish tussocks you can often find green shoots and frequently they are seen to have been nibbled by voles. The other grassland vole, the bank vole (*Clethrionomys glareolus*) distinguished by its much redder fur, especially in summer, and by its slightly longer tail, is extremely common in lowland valleys, sometimes riddling whole lane-side banks with its holes and runs. Not that the short-tailed vole is absent from the lowlands. In fact these two voles, with the long-tailed fieldmouse (*Apodemus sylvaticus*) often flourish as close neighbours in well-favoured, varied habitats such as sheltered gardens with rockeries, shrubberies and lawns.

Little is recorded about the occurrence of the yellow-necked mouse in Snowdonia. But as it is known from the lower Conway valley and reported from the lower Dovey valley, that is the northern and southern extremities of the Park, it presumably occurs at least locally in other places. It is a large, attractive-looking creature with a tendency to come indoors in winter, especially where apples are stored. But it is so much like the common long-tailed fieldmouse that it must often get overlooked. Any fieldmice caught in traps should be examined for a broad yellow band, not just a yellow spot, stretching across the breast between the forelegs. The Keeper of Zoology, National Museum of Wales, Cardiff, would be glad to receive specimens of the yellow-necked mouse with details of where they were found, in order that a more complete picture of this species' distribution can be built up.

The delightful dormouse seems to be entirely a low-lander, inhabiting wooded valleys in widely separated areas of Snowdonia. It is probably commonest in woods where there is thick ground cover and plenty of leaf-litter. It is well established in the Machynlleth-Aberdovey area, where it has several times been reported to have climbed up vegetation on a cottage wall to hibernate in a bedroom, on one occasion in a drawer full of socks. What could be cosier? More orthodox dormice hibernate in beds of dead leaves where, alas, they are sometimes scratched out by dogs. Dormice are far from being as nocturnal as some writers have said. I have seen them about in hedges by day and I watched one for half an hour one warm sunny afternoon climbing about the lower branches of a birch tree in a shady wood sucking the nectar from the tubes of honeysuckle flowers.

During some thirty years of this century (1920's to 1950's), while much of England was being overrun by American grey squirrels set at liberty in the south-east some years earlier, Snowdonia and west Wales remained a stronghold of the red squirrel and had no greys. But in the early 1950's when Pembrokeshire and Cardiganshire were invaded by grey squirrels it was clearly only a matter of time before Snowdonia would get them, at least in the south. So it was no great surprise when one was found killed on the road near Aberdovey in February 1955, and there were several sight-records the same year in Coed-y-brenin, Dolgellau, and one was shot near Penmaenpool that November. Since then the grey squirrel has continued to spread and is now widely distributed in the Park. For some years the Forestry Commission paid a small bounty for every grey squirrel killed, but as the irrepressible creatures continued to spread despite the large number that were being killed, the payment system was abandoned. Frequently the timing of the grey squirrel's infiltration into a new district followed a now familiar pattern. At first one or two very furtive ones were glimpsed here and there over a period of several months. Then there often followed a few months when none was seen and they were assumed to have died out. Then about a year after the pioneers were first reported they would be quite suddenly widely reported. Obviously this is a highly resourceful and adaptable species, for though a tree dweller it has been known to live down rabbit-holes and has been seen and trapped while crossing fairly high moorland several miles from any woodland.

It seems doubtful whether the newly arrived grey squirrels drive out the red squirrels in the combative way they are popularly supposed to do. Perhaps they successfully compete with them for food, especially in winter, or are the carriers of some unknown disease to which the reds are more susceptible than they themselves, for squirrels seem very prone to disease. Whatever the reason, almost wherever the grey invades the red declines; and this is now happening widely in the National Park. In many places where you often saw red squirrels they are now uncommon. I think the most likely red squirrel haunts now are the scrub-covered hillsides at the highest levels of deciduous woodland, especially in gorges and dingles;

or in high conifer plantations; for neither of these habitats seems to attract many grey squirrels. But red squirrels do not by any means abound in the conifer forests. In the early days of the Forestry Commission in Wales it was incorrectly forecast that red squirrels would multiply exceedingly in the new plantations. Perhaps they would have done so had the emphasis of the planting programme been on pine instead of spruce.

One matter particularly puzzles me about the red squirrel, or rather, about the literature on the red squirrel. I mean the statement more or less repeated by the best authorities that red squirrels' nests have no regular entrance and that the squirrels merely push their way in through weak places in the walls. Of the many red squirrels' nests I have examined all except one have had at least one carefully constructed entrance. The exception was a completely solid sphere that had neither entrance nor interior chamber and had evidently never been used. The best nests I have seen are those in which young were reared. They have had entrances, and sometimes exits, as definite as a dipper's entrance, though it is true that these holes have usually been at least partly concealed by an overhanging curtain of grass or moss. As for a squirrel just pushing in anywhere, this may be true of grey squirrels' nests, but the walls of most red squirrels' nests seem to me far too well knitted together to allow of it.

Some of the old naturalists were incurably what we would today call 'splitters'. They tried to split the marten into two species, the rock marten and the wood marten; some claimed there were two kinds of badger, the swine badger and the dog badger (from alleged differences in the shape of the head); and they made two or even three species out of the fox. Whether any tried to split the rabbit, I am uncertain; but many country people in upland Snowdonia are firm in their opinion that there are rock rabbits on the hills (they are alleged to be smaller, hardier and do not burrow), and bigger, less hardy, burrowing rabbits in the lowlands. Rock rabbits are said to be scarcer and widely scattered, whereas lowland rabbits live in large densely populated communities. These are no doubt differences of habitat not of species, but otherwise they are probably fairly valid distinctions and were important when myxomatosis first got to Snowdonia in 1954, for then most lowland warrens were almost wiped out while the upland rabbits remained disease-free in many districts. Since the decimation of 1954 rabbit numbers have remained small and wherever they do begin to build up, myxomatosis usually flares up again, presumably because of deliberate introduction, at least in some places. In Snowdonia the decline of the rabbit has not caused a decline in the buzzard as it has in south-west Wales, where both rabbits and buzzards were so abundant before 1954. Although there were large warrens in some coastal areas such as Morfa Harlech, there were always far fewer rabbits and buzzards in Snowdonia as a whole, and the buzzards have never depended on rabbits as exclusively as they did in the south-west. Rabbits in medieval times were a very valued source of meat and a warren was a precious local possession. It is recorded that in the thirteenth century every house in the commote of Estimaner in south-west Merioneth paid a penny towards the upkeep of the warren; and we also hear of a 'connygreen', or warren, at Towyn in the sixteenth century.

Brown hares are found throughout the Park. I have seen them up to about 1,500 feet but I expect they range higher at times. I have found them commonest on low-lying wet ground near the coast, but they are also frequent in deciduous woods and young conifer plantations where the deep grass of the initial stages of forestry is excellent cover for them and the leverets. A number of introductions of the mountain hare from both Ireland and Scotland were made by North Wales landowners in the nineteenth century in order to add to the sporting interest of the uplands. Bala, the Conway and Lledr valleys, Penmaen-mawr, Snowdon and Glyder are mentioned as localities at first inhabited by these introduced hares. They did well for many years and even today their descendants are still thinly distributed on high ground in Caernarvonshire, mainly on the Carneddau.

Deer, abundant in Snowdonia until the early seventeenth

century, were all gone by the early eighteenth at the latest. They were presumably steadily reduced by the clearing of their forest-haunts over a long period of time and then finished off by being hunted both for their venison and because they were a menace to crops. As Leland reported in the sixteenth century, there was little corn grown in upland Caernarvonshire: 'If there were the deer would destroy it.' That the large and conspicuous red deer should have been exterminated was to be expected. What is perhaps surprising is that the small and elusive roe deer did not survive here and there and that it has never recolonised Snowdonia since, not even in these days of large conifer forests. The fallow deer has perhaps never been wild in Snowdonia except as an occasional park escape. It is now the only deer and exists in one place only, the park at Nannau, near Dolgellau, a deer park that is almost certainly ancient. The increasing little muntjac deer which has spread across the English midlands so quickly and surreptitiously in the last few years is still very far from Snowdonia. But it has lately got into south-east Wales and may eventually work its way northwards on a similar route to that of the grey squirrel.

Reptiles and Amphibians

Though all the four species of reptile found in Snowdonia are widely distributed, only two, the common lizard and the slow-worm, are generally numerous while the other two, grass-snake and adder, are more local. The common lizard is particularly abundant and adaptable. It is found from the sand-dunes to the mountain tops; it is as characteristic of roadside banks in the lowlands as of the edges of mountain tracks; it likes dry grassy places, rocky slopes and walls; yet you can find it also (but less commonly) in quite wet sphagnum moss in bogs. Really watery places, the interior of woodlands and the sea-shore seem to be the only major habitats where you are not likely to see a lizard. Individuals vary much in approachability: most scuttle into the grass and are never seen again, but some you can inspect at close range and even pick up. Lizards sun-bathing, rival males fighting or pairs mating must be an easy prey for expert hunters such as kestrels and buzzards. Our other lizard, the slow-worm, only seems so much scarcer than the common lizard because it is mainly nocturnal and hides by day under stones or vegetation. It seems to be entirely a lowlander, being commonest in warmer habitats. It seems as plentiful near habitations and in gardens as anywhere and I have sometimes turned one up with the spade from several inches below ground. Like the common lizard, the slow-worm will occasionally emerge from hibernation in midwinter, at least in mild districts. There are January records of both from Aberdovey, the most southerly point in Snowdonia. Perhaps it is brightness of winter sun as much as warmth that brings them out. This might explain the emergence of those which are sometimes seen crawling over snow.

Two other lizards can be mentioned in connection with the Merioneth coast. A sand lizard was reported from Llwyngwril in August 1934 but has not been accepted as a good record. Large, well-marked male common lizards have too often been wishfully mistaken for sand lizards. All the same, remembering that the sand lizard used to be found on the North Wales coast near Rhyl, Prestatyn and Point of Air, I feel inclined to keep an open mind about the possibility that someone may yet find a sand lizard in the great dunes of Mochras or Harlech. Twenty specimens of the Continental green lizard, released in 1931 at Portmeirion by Sir Solly Zuckerman, survived a few years but evidently failed to breed. The grass-snake, with its preference for damp, even watery habitats, is naturally local. All the same it now seems more widespread than the adder. The present scarcity and patchy distribution of the adder is not easy to understand. All the evidence is that adders used to be greatly more abundant. H. E. Forrest's records (1907 and 1919) make the Snowdonia of his day appear to be comparatively crawling with them. Today they are particularly local in Merioneth but somewhat more widespread in Caernarvonshire where there seem to be a fair number in the Capel Curig area, and I have seen a few in the

Nantmor valley. Adder numbers are variable and they can become very abundant for short periods. In Merioneth in August 1936 there was such a plague of adders in the Artro valley near Llanbedr that workmen there were insured against snake-bite. Adders are not strictly confined to the dry, sunny banks and heaths usually thought of as typical adder country: as with lizards, you can find them in peat bogs, sometimes lying in wet patches of sphagnum moss. Some people seem to have such a deep urge to kill every adder they meet that I suppose it is useless to expect them ever to grow out of it. But surely at least they could learn to recognise grass-snakes and slow-worms and allow these harmless creatures to live? It always astonishes me that so many schoolchildren, even in these days of popular nature study, are apparently left in ignorance of these simple distinctions. Too often one finds a grass-snake or a slow-worm beaten to death at the wayside.

Perhaps the most interesting thing to find out about frogs in a mountain region is the height to which they range. In Snowdonia they breed at some of the highest peat pools, often when there is still snow on the mountains and even ice partly covering their pools. Because of the low temperatures and poor feeding in some of these acid mountain pools, tadpoles may be very slow to develop. Some of them take at least a year and grow bigger than normal tadpoles. Both spawn and frogs are much sought after by birds and mammals. I have seen crows, buzzards and herons taking this easy prey, and others have reported foxes, polecats and stoats as eating mountain frogs. I once found a pile of frogspawn in a buzzard's nest in a cliff, but that may have been part of the decoration buzzards like to put on their nests early in the season. Especially in spring, but also at other seasons, a whitish jelly-like substance not unlike frogspawn can be picked up on wet moorland. If you ask an old shepherd what it is he will probably laugh and tell you he doesn't know but it's what his father used to call *pydru ser*—the rot of the stars, a fanciful and probably ancient attempt to explain what happens to a shooting star when it falls out of the sky.

In an article in *Nature* in 1926, H. A. Baylis reported that he had examined this 'rot of the stars' or 'star-slime' and found it to consist of 'parts of the viscera of either frogs or toads' presumably the victims of predatory animals or birds. 'What appears to happen is that the gelatinous secretion of the glands lining the oviducts splits open and their contents soon assume the appearance of an amorphous jelly.' On the other hand, botanists who have investigated *pydru ser* claim it as a fungus. So if both zoologists and botanists are right, *pydru ser* is a name for at least two very different substances closely resembling each other.

The common toad is very abundant in damp or shaded places throughout lowland Snowdonia but is apparently rare or absent on the high ground. In John Williams's day (the 1820's) there were natterjack toads at Llanrwst, where they were again reported about the end of the century but not since.

Of the three British newts only one, the palmate, is found throughout Snowdonia, for it is the only one recorded from the high ground; which is appropriate enough since its other name is alpine newt. The distribution of the other two newts seems to have had little attention. Both are very locally recorded but the details of their presence or absence over wide areas are quite unknown. It may well be that both the crested and the smooth newts are rare in the Park. The only record Forrest could find of the smooth newt was from the Little Orme. An investigator in the 1930's claimed to find both palmate and smooth newts in the Aberdovey district.

Fish

To judge by what you hear from some anglers (and all poachers) there are only two fish in Snowdonia: the salmon and the sewin (or sea-trout). It is an understandable attitude for they are magnificent fish. It would be splendid to observe them in the sea, which hardly anyone ever does, but to see them inland far up the Dovey, the

Mawddach, the Conway or the many other rivers, whether as dark and massive shapes lying in quite small pools or as silvery splashing monsters leaping at the falls, is an exciting experience. The salmon is always at the centre of controversy, sometimes real trouble. Scientists argue with each other and with anglers over its natural history. For instance: do salmon eat or not as they ascend to breed, and if not why do they rise to a lure? There is a cold war between upstream anglers and the estuary netsmen who argue over their respective rights; and sometimes a hot war between water bailiffs and poachers. And knowing nothing of these matters until they meet either net or hook, the salmon live out their mysteriously alternating sea-lives and river-lives and go through their remarkable transformations and migrations, finding their way by senses which are fantastically acute.

Then there are the members of the family that do not migrate to the sea and back but make much shorter moves to reach their breeding-grounds: the trout, the char, the gwyniad and the grayling. The trout is by far the commonest fish of Snowdonia. It inhabits all rivers and lakes except where there is serious pollution (which mercifully is very rare) or where the water is excessively peaty, as in some small shallow moorland lakes. It grows rapidly in some rivers and a few of the lakes: but angling pressure is now so heavy except in a few upland lakes difficult to reach that the chances of any trout or salmon reaching great size must be increasingly slight. In some rapid, bouldery streams where feeding is poor, the trout do not grow to more than a few inches in length no matter how long they live. The beautiful but little-known char, formerly a little more widespread, is now known in only four lakes: Peris, Padarn and Cwellyn; and Bodlyn near Barmouth. They are deep-water fish rarely observable at the surface. They come into the shallows near the lake-edge or up streams near the lake for spawning in November.

Like the char, the gwyniad, unique among British fish for having a Welsh name only, was often mentioned by early travellers. But it was famous long before the travelling era. The *Description of Wales* of 1599, as I mentioned in chapter 1, makes an early reference to this fish in Bala Lake. By 1695 there had been an advancement of learning and we find Lhuyd telling us: 'The word gwyniad may be aptly rendered in English a whiting but is very different, being of the trout kind. A description of it may be seen in Mr. Willughby's *Ichthyology* who supposes it the same with that they call ein Albelen and Weissfisch in some parts of Switzerland, and the Ferra of the Lake of Geneva. They are never taken by any bait, but in nets; keeping on the bottom of the lake and feeding on small shells and the leaves of water gladiol, a plant peculiar to these mountain lakes' (and which we now call water lobelia). The gwyniad may be unique, but it is very closely related not only to the Swiss species known to Willughby but also to the several whitefish of Scotland, Ireland and the Lake District (variously called vendace, pollan, powan and schelly). Because allied species in North America are migratory it is thought that the gwyniad, and the Welsh char too, were perhaps also migratory a very long time ago. The only other fresh-water member of the salmon family in Snowdonia is the grayling, which is absent from Snowdonia except in Bala Lake and presumably at least a little way up the Dee above the lake.

The scarcity of the grayling is but one example of the most striking aspect of Snowdonia's fish: the absence or rarity of many species found commonly in English rivers even as far west as the Welsh marches. Besides the grayling one could mention roach, dace, rudd, chub, pike, bream, gudgeon, bleak, miller's thumb, stone loach and others, some of which are found in Bala Lake but are absent or rare in the rest of Snowdonia. Some of the absentees come closest to the Park boundary in the Dee below Bala, the Dee being the one river that flows eastwards from the Park. Bolam found the loach in the Lliw stream that flows into the Dee near the head of Bala Lake, and this seems to be the only loach record for the Park. Bolam also records the gudgeon as being in the Lliw and its neighbour, the Llafar, and they are also the only localities

I have found mentioned for this fish in Snowdonia. The bullhead, or miller's thumb, is recorded from Bala (again the Lliw), Bangor and Llanbedr, and is thus very widely distributed yet narrowly restricted to streams round the circumference of Snowdonia. Though for centuries of the Middle Ages there were religious houses in Snowdonia (at Beddgelert and Llanelltyd for example), they have left no carp behind them. If they kept carp it must have been in stew-ponds long since destroyed. There are perch in several of the Snowdonian lakes, some of the highest being in Llyn Arennig Fach at nearly 1,500 feet. They are also very common in Bala Lake; and in a few others where possibly they are an introduction, as the pike is also.

Sticklebacks are by no means as common in Snowdonia as they are in parts of England. The typical upland trout stream usually has trout and salmon only and certainly no sticklebacks. Although in the lowlands sticklebacks are common in ditches and pools and also in the brackish water that lies in the hollows in salt-marshes, this species is apparently quite absent from the mountain region. Not so the minnow, which does very well in quite a number of upland lakes. Forrest had no record of the ten-spined stickleback in Snowdonia, but that versatile naturalist Mary Richards tells me that as a child she used to catch them in brackish ditches on Morfa Dyffryn. That would be about the beginning of this century, the same time, in fact, as Forrest was collecting his records.

A complete list of estuary fish would be a long one for it would include many sea-fish that the tides bring in occasionally. I will mention only those which are both common and regular. After salmon and sewin, surely the best known are the eels which, as wriggling black masses of elvers, come out of the sea and up the rivers and streams in spring. And surely the best-known eels are those that wriggle up the rocks to get round the falls in late May in Major Lloyd Jones's garden at Betws Garmon on the north-western edge of the Park for they are seen by many visitors. This is a private garden but the public are invited to pass through, which they are reported to do in numbers but without doing any damage or leaving litter—a splendid example of the sort of co-operation between landowners and public one would like to see operating throughout every National Park. Adult eels, and perhaps congers too, are a favourite prey of cormorants, especially in the estuaries where the birds can often be seen with large eels writhing in their beaks before eventually being swallowed.

Also conspicuous in the estuaries are the grey mullet which swim in shoals near or at the surface, sometimes with little silvery fish leaping out of the way in front of them. These little leapers are probably young herrings, which the fishermen call 'britt' or 'whitebait'. They often come in large numbers and are followed by many other predators, especially fish such as mackerel and birds like gulls, terns and mergansers. The adult herrings have declined amazingly in North Wales waters since the days when there was a herring industry here. For instance, in the state papers of Elizabeth Tudor's reign there is a survey of the Merioneth ports dated 1565 which says of Aberdovey: 'being a haven and having no habitation but only three houses whereunto there is no resort, save only in the time of the herring fishing, at which time of fishing there is a wonderful great resort of fishers assembled from all places within the realm, with ships, boats and vessels; and during their abiding there, there is of the said company there assembled one chosen among themselves to be their admiral.' I imagine this gathering as a sort of August holiday when inland people crowded to the sea and got all the herrings they could for pickling, for this fish has long been a valued item of winter food. Bass, flounder, smelt and garfish are well-known estuary fish. The eel-shaped garfish are particularly beautiful: when you can watch them from above swimming near the surface you can see wonderfully iridescent blues and greens along their backs. Sea-lampreys, detested by fishermen for their ability to gouge large holes out of the flanks of big salmon, are common in the estuaries and have been taken over three feet long in the Dovey. They ascend the rivers to breed and have been recorded more than twenty miles up the

Dovey nearly at Dinas Mawddwy where, according to Forrest, in the late 1870's they were caught 'in such numbers that they were used to feed the pigs'.

Insects

I do not know how many of the 20,000 British species of insects are to be found in the Snowdonia National Park. I do know that in this short section I am going to mention only the very few that attract most attention: I mean the butterflies, a handful of moths and a round-up of miscellaneous other insects. If I leave out whole orders of insects and indeed whole classes of animals, I justify such omissions on the grounds that specialists in any group will already be in touch with the relevant literature.

Butterflies first: forty-two species are recorded for the Park area. I mention only the more interesting here and give a complete list as an appendix.

The brimstone has always been rare because its food-plants, the two buckthorns, are also rare. In fact the common buckthorn (*Rhamnus cathartica*) is probably entirely absent and the alder buckthorn (*Frangula alnus*) is very local. The brimstone is probably least rare in the lower Dovey area because the alder buckthorn is not uncommon there. This fine butterfly can also be expected (but as far as I know has not been recorded) in the Ganllwyd district, the best buckthorn area in Merioneth. As the Forestry Commission have here and there planted alder buckthorn to beautify their plantations there is every likelihood of the brimstone increasing. Six species of fritillary are widespread but all are restricted to suitable habitats. The dark-green is probably the most often noticed for it is often abundant from the sand-dunes to the lower hills. Both the two pearl-bordereds are locally plentiful, especially, I think, the small pearl-bordered which frequents bogs, marshes and damp woodlands from sea-level to the moorland valleys. The silver-washed occurs in wooded districts, often on bramble flowers, usually in late July or early August but some years even in September. The high brown seems to me the scarcest of the Snowdonian fritillaries, for though the marsh fritillary has the reputation of rarity it is abundant in several upland and lowland localities, whereas the high brown, though perhaps more widespread, is very much more elusive.

The comma, formerly a rarity, has increased and spread greatly this century and is now not uncommon. It is usually seen in the lowlands but I have seen it at 1,000 feet and others record it higher still. To offset the increase in the comma we have to accept that the large tortoiseshell, always very rare, has probably quite gone. It is still said to occur in the country round the Conway estuary, but not within the Park. Elsewhere there have been two interesting reports of it by independent observers, one at Llanbedr on 17 July 1945 and the other at Aberdovey on 13 August of the same year. These two places are both on the coast about twenty miles apart.

Speckled woods and graylings are both common. The grayling is more abundant towards the edges of the Park than in the interior, especially in sand-dunes and on hills by the sea. One of the commonest species is the small heath, for in a predominantly mountain region this is the only butterfly that is resident from the lowlands to really high ground. It is in fact the only truly mountain butterfly of Snowdonia. The large heath gets nearly as high—on the Berwyn, for instance, but is rather a rarity of upland and coastal bogs.

The Duke of Burgundy is evidently very local and I have never seen it. It has been recorded in Merioneth from Arthog, Dolgellau, Llanbedr and Tan-y-bwlch. Of our four hairstreaks only the green is easy to find every spring, for it flies about grassland from sea-level up to the edge of the moors. In the hillside oakwoods the purple hairstreak is widely distributed but varies a great deal in numbers from year to year. Both the brown and the white-letter hairstreaks are extremely rare: the brown is recorded (but not recently) from Trefriw in Caernarvonshire, and from Aberdovey and Bala in Merioneth. The white-letter was recorded from the Aber valley, Caernarvonshire, in 1942; but in Merioneth not since 1904, at Barmouth. I give these old records because in areas where comparatively little collecting goes on it is always possible that a rare species may have persisted quite unnoticed.

The holly blue is apparently very restricted in Snowdonia but is locally plentiful. There are a number of records from the Conway valley up to Betws-y-coed and it is frequent some years at Aberdovey and Llanbedr. Like so many of the Park's butterflies, it is probably rare away from coastal or estuarine districts, but there is a record from Beddgelert. There are old records of silver-studded blues at Harlech and Barmouth—curiously isolated occurrences. Of our four skippers the large is easily the commonest; the dingy is local; and the grizzled seems decidedly uncommon, the eastern and southern parts of the Park being the most likely places for it. The small skipper is locally frequent in the south of the Park.

I limit what I have to say of the moths mostly to upland species to be found above about 1,000 feet. The largest and most conspicuous on the open moors are emperor, fox, northern eggar and drinker. Their caterpillars are usually far more conspicuous than the moths, especially the fox, a hairy caterpillar with black and yellow rings when small which become black and red-brown as it grows. The yellow-spotted, brilliant green caterpillars of the emperor moth are the most splendid of creatures. You usually come upon them one at a time, but sometimes a patch of bilberry or heather will be crawling with them, especially when they are young. They are then in various patterns of black, orange and green.

Other attractive species of the moorland are ruby tiger, wood tiger, clouded buff, northern rustic, true lovers' knot, beautiful yellow underwing (truly beautiful and often flying by day visiting heather flowers), antler (caterpillars superabundant in mountain grassland in some years), Haworth's minor, green carpet (a common lowland moth that occurs high up in mountain cliffs also), grey mountain carpet (abundant among rocks on the highest mountains), northern spinach, July high-flier, common heath (one of the most numerous of moorland moths), and the forester. Caterpillars of some interesting species such as poplar hawk, eyed hawk, certain prominents and kittens, puss moth and a number of others can be found on willows and aspens as high into the mountains as these trees grow. Ashworth's rustic is a beautiful moth restricted to North Wales but apparently not uncommon here, especially in the higher valleys. I have seen its unusually marked caterpillar (slate-grey with two rows of heavy black oblong spots) in Cwm Cowarch near Dinas Mawddwy. Lowland species of particular interest are the scarlet tiger, a gaudy moth that flutters weakly over bogs in June and, though generally rare, is locally common; and the alder, whose caterpillar is occasionally found along hedges and woodsides and looks like a bird-dropping when small but when larger is ringed yellow and black and has remarkable long club-ended hairs standing out along its sides. Both the broad-bordered and narrow-bordered bee-hawk moths are found in Snowdonia, mainly in coastal districts; neither is common, but the narrow-bordered has been recorded several times at Aberdovey. A bee-hawk hovered about me for several minutes one July morning on the top of Pen-y-gaer hill-fort in the Conway valley. A very special little North Wales moth is the weaver's wave, which is found both in the Park and just outside its northern edges, and nowhere else in Britain.

I conclude with a few other insects of general interest. Dragon-flies are numerous in both lowland and moorland habitats, the big golden-ringed dragon-fly being the most conspicuous upland species. Beetles may look a formidable study with nearly 4,000 species on the British list, but for encouragement I point out that Dr. J. H. Salter of Aberystwyth did not take up the study of beetles until he was in his seventies and went on to become an expert. A conspicuous beetle in early spring is the bright green tiger which flies a few feet at a time in front of you along the hill tracks. Then in early summer, soon after the cockchafers as a rule, that smaller chafer, the June bug, hatches out of the turf in vast numbers. It is reddish in the body with a metallic green head and is swallowed in great numbers by jackdaws. Click beetles too are enormously common in some years in the mountain pastures. That strangest of beetles the glow-worm is locally frequent in the valleys. Large bumble-bees appear very early on high ground as soon as winter is over; and in summer wild bees of various kinds are abundant in the hills, especially in the heather, as was duly noted in Lewis's *Topographical Dictionary*: 'the large heathy mountains, more particularly those of the Berwyn range, swarm with a species of wild bees, and on a fine day these wilds may be traversed for miles without hearing the least noise save the monotonous hum of these busy insects'. That largest of ants, the wood ant, which makes huge piled-up nests, is found in many lowland and semi-upland woodlands of the Park, especially I think in the northern and western areas, but it occurs near Bala also. Flying ants, usually in August, attract spectacular, acrobatic flights of black-headed gulls and other birds. That striking but harmless insect, the giant wood wasp (*Sirex gigas*) which sometimes blunders past you with a heavy roar of wings is quite common in Snowdonia and as elsewhere is preyed on by the large ichneumon fly (*Rhyssa persuasoria*), which is just as skilful a wood-borer as its victim. Crane-flies, or daddy-longlegs, especially smaller species, can be seen hatching in great numbers in the moorland turf in spring and summer. A winter-gnat (*Trichocera maculipennis*) is abundant nearly a mile inside a slate-mine at Corris. A rather large, sluggish, black fly is conspicuous in August, especially on wet heather moors. It has long, dangling legs and a long thin body and looks very much like the St. Mark's fly of spring but is a different species named *Bibio pomonae*. It is often seen in great numbers crawling over the heather, and, being a weak flier, often gets blown on to the surface of lakes and is avidly taken by trout. Finally, the pestiferous horse-flies which are at their worst in July. There are over two dozen species but happily only a few are troublesome to man. The commonest and most insidious is the small grey species, *Haematopota pluvialis*. Also abundant and most persistent is a mottle-winged species with extra-brilliant green eyes, *Chrysops caecutiens*. And, quite formidably huge, is *Tabanus sudeticus*, with a nearly two-inch wing-span. Fortunately, though it buzzes round you in a rather terrifying way, it is much slower to settle and bite than the others and is more easily discouraged. Mountaineers will know that many of these blood-sucking diptera are also mountaineers!

CHAPTER FIVE

The Mountain Birds

In 1797, when the Welsh travelogue frenzy was at its height, someone wrote: 'Scarcely a summer passes but the opulent or the curious from the most distant parts visit the principality and volume upon volume is written to record its minutest beauties.' How I wish these last few words were true! For then, instead of the same old route being followed, the same old descriptions of the Lake of the Three Pebbles, the Mawddach waterfalls, Aberglaslyn Pass and Llanberis Lake, we might really have been given some account of the region's 'minutest beauties' in the shape of the plants, insects, reptiles, mammals or birds. As it is, apart from a few plant-lists, some of doubtful authenticity, these writers give the naturalist practically nothing. If only the inquiring spirit of Lhuyd had been carried over into the next two centuries! But in the Wales of his day, and in fact until very recently, public acclaim went not to scientists, but to preachers, poets and politicians. The result is that we know practically nothing about even the fairly recent history of most of the animals and birds of Snowdonia. To think of all those writers of the 1790's, wandering about Wales with time on their hands and looking desperately for something new to write about, yet never giving a glance, most of them, at the colourful world of living nature vibrating so obviously about them! What I have to say about the birds of the Park would have added interest had I been able to make comparisons with accounts written 150–200 years ago, for I could then have shown how this or that species has fared with the passage of time. Bird populations, like all else in nature, are not static. Natural changes in number, distribution and even habits are going on all the time, some rapid enough for all of us to see but most too slowly for any of us to realise unless we have access to information stretching over centuries.

It seems most appropriate to describe the birds of a mountain region by beginning at the top. If so, the first bird I must mention is the chough, for in my personal list of Snowdonia birds the chough holds the height record on the strength of two I saw one late summer day when I was on Lliwedd (2,900 ft.). I would certainly not have seen them if they had not been calling, for they were no more than two specks against a white cloud, flying westwards. The only other birds you are likely to see high up over the highest peaks are ravens at any time of the year, and swifts which come arrowing over the tops on fine summer days though their nests are miles away down in the lowlands.

Ravens are surely the hardiest of the mountain birds. Even in very severe weather when all other birds have retreated off the tops you are quite likely to hear the bark of a raven some time during a day's walk, for it is a sound that carries far on still, frosty air. Even in the arctic winters of 1947 and 1963 the ravens nested in February as usual. A long, snowy winter means extra numbers of dead sheep lying about, followed by further casualties in spring when the weakened ewes die in giving birth. Not that the raven is at all dependent on carrion in the way some vultures are abroad. Ravens are highly adaptable as to diet and greatly skilled at finding food when food is scarce. What affects them most is persecution, for their large size and appetites have made them such a special target for the hostility of farmers and gamekeepers that there are now wide areas of central and western Europe, and of England, where there have long been no ravens at all. And by 1914 they were becoming rather uncommon in much of lowland Wales. No doubt the remoteness of some of their Snowdonian haunts saved the ravens here. From these mountain strongholds they were able to spread back down the valleys when the pressure of persecution was suddenly taken off by the 1914–18 war which diverted gamekeepers to other occupations. After that war gamekeeping was never resumed on the pre-war scale in most of Snowdonia and the raven has become quite numerous again in uplands and lowlands alike. Perhaps nothing illustrates the increase of the raven more clearly than a comparison of flock sizes then and now. H. E. Forrest records a dozen ravens together in 1913 as 'an extraordinary gathering'. But in recent years flocks of 40 or 50 have become quite normal, especially at a regular all-the-year-round raven roost in south Merioneth.

It is as ravens come in to such a communal roost that you get some idea of the strange noises they can produce. Besides the normal *kronk*, they pour out a wide range of squeaks, trills, growls and explosive pops. Then there is a falsetto *prork*; and a deep, resonant *gerronk* which sounds far across the hills. But the most curious note of all, which I have heard mostly but not always in spring, is like the sound of two stones dropped quickly one after the other into an empty bucket. Many of these notes accompany the aerial tricks that identify ravens at great range, the commonest of which is to turn upside down in mid-air. Usually the bird rights itself immediately but occasionally one will glide on for several yards breast uppermost.

The great size of ravens is often not apparent in the field because of the vastness of their surroundings; but their shape and flight are distinctive. No other crow soars so much. I have put ravens up off dead sheep and seen them, reluctant to leave the place, soar round and round overhead, gaining height until almost out of sight and with never a wing-flap the whole time. When soaring, a raven, with long straight wings, long neck and well developed tail, looks like a black cross in the sky. The wings have great power and in full flight are noisy, making a loud *whoo-whoo-whoo-whoo* that is sufficient for identification even without the bird being seen, as, for instance, in thick mist. The extra length of a raven's neck is not without its use for it allows to the throat an uncommon degree of distension for the pouching of food which may have to be brought long distances. A raven carrying a full load of food to its young can be told afar off, so conspicuous is the bagged throat, made even larger by the shaggy pointed feathers bristling on it. Some ravens' nests, added to year after year, grow into stacks six feet high, and in May, when the grey-eyed young are full grown, such nests can be visible a vast distance, plastered white with droppings and conspicuous against the dark grey rocks.

One naturally thinks of the peregrine falcon in connection with the raven, for the two often breed on the same stretch of cliffs though usually keeping their nests at least a few hundred yards apart. Old ravens' nests clearly exercise a fascination over peregrines, which often rear their young in them or at least use these great piles of sticks as a roost, a look-out or a mating platform. There is frequent trouble between ravens and peregrines at nesting time for the falcons, ever watchful for their eggs and their young, harry the ravens constantly. You may have admired the deftness with which a buzzard turns over in mid-air to avoid the stoop of a crow. But that is slow motion compared with the game played by raven and peregrine. The perpendicular stoop of the peregrine is almost too quick for the human eye. But the raven always manages that last split-second twist that enables him to point his great beak upwards and so keep off the falcon as she passes with a roar of vibrating feathers. What is wonderful is how the falcon then turns at the bottom of her stoop and comes back vertically upwards with no apparent loss of speed to attack the raven again from below. Again he evades her as she shoots up past him, turns and stoops again. So it goes on all the way across the valley until the raven reaches the safety of his ledge. I suppose most of the peregrine's aggressiveness is threat-display rather than serious intent to kill. But I wonder, if the raven for once took no evasive action, whether the falcon would really strike?

Alas, when I speak of peregrines and their ways I speak of what could have been observed easily a few years ago but which has now become a rare sight. For lately this magnificent bird has suffered a very severe decrease in numbers. There is every reason to attribute this disaster to the use of poisonous seed-dressings. The dressed seed is picked up by grain-eating birds which, in a semi-poisoned state, are taken by peregrines, perhaps mostly in autumn and winter when they range widely away from the mountains. Eventually the peregrines accumulate enough poison in their bodies either to kill them or leave them barren. In many countries, a campaign is being fought to get the use of poisons on the land prohibited or greatly restricted. All people sympathetic to wild life await the outcome of this campaign with anxiety.

So far the kestrels of the uplands have not suffered the fate of the peregrine. The kestrel feeds very differently. In the mountains its food is mainly voles and insects and if it occasionally takes birds they will be pipits, larks or wheatears, insect-eaters of the mountains and unlikely to be carrying poison in their bodies. So on the hills the kestrel remains widespread and fairly numerous although it has become rare in parts of lowland Britain. This small falcon is to the crow what the peregrine is to the raven: a tormentor and a rival over nest-sites. Just as the great crags are disputed between ravens and peregrines, the smaller rocks are often the scene of ferocious rivalry between crows and kestrels. Just as old ravens' nests attract the peregrine, so old crows' nests attract the kestrel. And sometimes kestrels take possession of crows' nests actually in use. This is achieved by a process of persistent harassing. One pair of kestrels I watched took up a position on the top of a crag where crows were completing a nest on a ledge half-way down. Every time a crow left the nest or returned, one or both kestrels dropped like an arrow just past the crow, often causing it to tumble in the air. Then, exactly as the peregrine, the kestrel would turn and come back vertically to skim up past the crow from below. This went on hour after hour for two days, by which time the crows had had enough. The kestrels then went to the crows' nest, tore out all the wool lining and left it hanging conspicuously over the edge of the rock. A fortnight later I found their four red eggs where the green eggs of the crows would have been.

I suppose if you measure success by large numbers, wide range and great adaptability, it could be claimed that the carrion crow, which is just as at home in a town park as on a mountain crag, is the most successful of all our birds. Success breeds unpopularity and the crow is certainly unpopular, not only among farmers and gamekeepers but even among naturalists, many of whom reckon it unhealthy for the natural balance of life that so large and hungry a predator on smaller birds should be so numerous. Crows, it seems reasonable to argue, were less common in the days before man wiped out or reduced the buzzards, kites and goshawks which used to prey on them. The shepherd's horror of crows is understandable. A very weak lamb or a helpless ewe is quite likely to have its eyes pecked out by crows or ravens. When that happens the argument that most of these sheep would have died anyway and that the crow is for most of the year quite an innocent feeder is not likely to impress the man who has to put blinded animals out of their misery.

The crow is abundant throughout the Park except on the highest crags, which are the domain of raven, peregrine and sometimes buzzard. All the same, there is no shortage of crows anywhere on the high ground because they forage everywhere at times. In trees they build very skilfully, binding their nests together as tight as lobster-pots. But on mountain ledges their nests can be quite slipshod, nothing more than a wool-lined circle of loose sticks round a scrape.

If these four birds—raven, peregrine, kestrel and carrion crow—are fairly evenly distributed in the Park, the same cannot be said of the buzzard, the only other large bird of the crags. Buzzards are decidedly commoner in the south, in Merioneth, than in the north, in Caernarvonshire. Like the raven, the buzzard was getting scarce through persecution at the beginning of this century. But since 1916 it has made great advances. Though it suffered a considerable set-back in some lowland areas of Britain from 1954 onwards when myxomatosis decimated the rabbits, its numbers in mountain areas such as the Snowdonia Park were far less affected because rabbits there were always less common and therefore of less importance in the mountain buzzard's diet. It is regrettable that, despite its being a protected bird, the buzzard is still being killed by a few farmers and gamekeepers, for this is a very fine bird to see floating and calling about the sky and is sufficiently eagle-like to give a touch of real wildness to its haunts. And people who do not care whether they see buzzards or not should be informed that the number of young grouse or poultry chicks taken by buzzards is probably very small indeed compared with the number of crows and magpies they kill. I have seen many a buzzard's nest, especially in July when the young are big, quite heaped up with the remains of crows, jackdaws, rooks and magpies, while other carcases lay on the ground below. But keepers and farmers rarely bother to investigate this side of the buzzard's life.

Though buzzards are quite able to meet the competition of ravens and peregrines and breed near them on open mountain cliffs, they mostly prefer to hide their nests in lower heathery terraces such as you often see on the sides of gorges and waterfalls. Their staple food in the uplands is probably that of the kestrel, namely the short-tailed vole that is so abundant in mountain grasslands. But, as I have said, the young buzzards are very frequently fed on young of the crow family, young rooks being an obvious and easy prey when they come foraging in the hills in late spring and summer. But the buzzard, though a characteristic mountain bird, is—like the carrion crow—more abundant down in the wooded valleys.

There remains one more upland bird of prey, the merlin, by far the most elusive of them all and now undeniably rather scarce. All the late-nineteenth-century records show that merlins were more plentiful in those days. Since then they seem to have steadily, unaccountably declined, though their principal food, the meadow pipit, remains abundant. But though scarcer, there are still merlins throughout Snowdonia, mainly on the heather-moors and on rocky, heathery mountainsides in the breeding season and on coasts and estuaries in winter. A few also breed in grassy areas of the lower hills far from any heather. There, in the absence of thick ground cover, they lay their eggs in old crows' or magpies' nests built in small isolated trees in or just above the topmost fringe of the deciduous woods. Merlins are a delight to see at any time, flickering quickly past you in the hills or twisting and fluttering round some dashing emperor moth (which usually seems to escape) or simply playing high in the air as they sometimes do in early spring. Everything merlins do seems to be done with such dash and fire. But they are at their most spectacular in their courtship pursuit in April when the male pursues the hen at fantastic speed down the hillsides, skimming low over the ground, twisting madly round rocks, both of them shrieking wildly.

After the fugitive merlin it is a relief to think that there is one bird that really proclaims his presence among the rocks and the heather: the cock ring ouzel. From early April to July you may hear his note, three or four times repeated, in many of the wildest places in the Park. For his song, like that of the golden plover, is almost exclusively a mountain sound, though occasionally you may hear it from ring ouzels on passage up the coast. It has a strange quality, this simplest of songs that carries far in the stillness of the hills. And it can sound quite eerie and unbirdlike when it comes down from the black shadows of a crag at nightfall. Though especially characteristic of tall heather whether it grows on craggy slopes or peaty moorland, the ring ouzel also frequents quite different habitats such as old slate-quarries, abandoned mines and deserted moorland houses far from heather.

Although they often advertise themselves by song or by chattering alarm cries, ring ouzels are extremely watchful and can be very secretive, slipping quietly round the rocks into the next gully when you are still a long way off. They are very agile and fast over short distances and can fly down a slope at breakneck speed to drive off a crow or a buzzard and will pursue them, screeching like a mistle thrush, for several hundred yards. Their nests and eggs, usually well hidden in heather but sometimes on ledges in buildings or mine-shafts, are very like those of the blackbird. So are many of their feeding habits. In spring you can see them pulling worms out of short mountain turf like blackbirds on a suburban lawn. Then later, as blackbirds do, they turn to fruit: in July the bilberries, in August and September the berries of the mountain ash. Then off they go to their wintering grounds in the Mediterranean region. Most of them return to the hills (sometimes in flocks of a dozen or more) at the end of March or very early in April but there are much earlier records. Bolam, for instance, saw five near the top of Aran Fawddwy on 19 February 1905. The ring ouzel was one of the few birds to catch the attention of several of the earlier writers. Lhuyd mentions seeing it in both Caernarvonshire and Merioneth (Camden's *Britannia*, 1695) and the Reverend Bingley reported it on Snowdon in 1798: 'Among the higher rocks I observed the Black Ouzel, *Turdus torquatus* of Linnaeus: it is not an unusual inhabitant of these alpine regions.'

Mountaineers and rock-climbers not particularly interested in birds might well fail to notice the unobtrusive notes of the ring ouzel. But no one could miss the song of the wren. Wren song is loud enough even in a lowland wood full of other singing birds, but it is really startling when it suddenly bursts out among the crags and buttresses of, say, Cwm Idwal or the north face of Cader Idris at nearly 3,000 feet. But the most striking quality of wrens is their ability to withstand the mountain winter, for you can find them still holding out among the high crags when all the world is frozen and nearly all other birds have retreated to the valleys. This survival of mountain wrens is particularly remarkable when we remember that wrens elsewhere are not usually considered hardy, for they perished in great numbers in the lowlands in severe winters such as those of 1947 and 1963, though surviving in the mountains. I think the explanation of the survival of mountain wrens must lie in their cave-dwelling habits. In a hard winter it is the deep snow-cover and the strong, persistent, shrivelling east wind that kills. Those plants and animals that can keep down in damp cavities under the rocks can enjoy a humid local climate no matter how searing the wind is outside. The water that seeps down inside these sheltered crevices never freezes and so in them you can find long fronds of ferns looking perfectly fresh and green when all fern-fronds exposed to the east wind have long been brown and dead. If ferns can survive in these places it seems reasonable to suppose that this is where the wrens winter too, finding enough spiders, woodlice or other small creatures stirring to keep them going till spring. It was certainly wise of Linnaeus or whoever it was who first called the wren *Troglodytes*—the cave-dweller, for it is evidently a bird more fitted to survive in cave conditions, even in mountains, than in the woods or gardens of the lowlands.

A bird that in the nesting season often shares big bouldery screes with the wren but is certainly no cave-dweller is the wheatear. But though wheatears are numerous and breed from sea-level to the mountain tops they are not everywhere in the mountains. Even on ground between 800 and 2,000 feet, which is their stronghold, there are innumerable small gaps in their distribution in Snowdonia. They tend to be grouped in local, loosely-knit communities. Where you find one breeding pair you are likely to find several others nearby. It is the same with whinchats, ring ouzels, curlews, sandpipers, dunlins, golden plovers and perhaps most species except the superabundant.

Wheatears belong most characteristically to stretches of short turf scattered with boulders: short turf because they can run and dart quickly over it after insects; boulders because they need them as vantage points and singing posts. Clearly the mountain sheep, keeping large areas of grassland closely nibbled, is an essential ally of the wheatear. The same could be said of the mountain wind that keeps the summits almost bare of vegetation, for the wheatear is one of the very few birds that can find a living up there, using the summit cairns for sheltering, nesting and perching. But summit-dwelling wheatears are not common. Where wheatears are commonest is perhaps along the wall-margined sides of moorland tracks near the highest occupied farms, tracks that are in frequent use by farm animals whose droppings attract a concentration of insects. There the wheatear will sometimes nest in the walls but more often in a hole in the ground under a rock.

The first wheatear of spring can be a delightful surprise. It is mid-March and in the mountains still quite wintry. Neither larks nor pipits have yet moved up from the valleys and the world is looking pretty lifeless when, perhaps from behind a snowdrift, a cock wheatear unexpectedly appears, looking very alert and bright in his breeding plumage. From now on all through spring the hovering courtship displays and plungings of the cock wheatears and their squeaky, scratchy songs will be a welcome touch of life among the grey rocks. After mid-June the young will be out, flitting about screes and roadsides, conspicuous with their very white tails. In a good summer a second brood follows the first and by September the wheatear families can be very plentiful. But by then all wheatears can look rather alike in the field because most of the males at that season have moulted their breeding dress and turned as brown as the females and the young.

When carrying food the adult wheatears usually stand and watch you warily as you approach. But even more distrustful are the meadow pipits. They are extremely reluctant to take food to their young if you brazenly try to watch them back to the nest. They will play out the patience game standing on a rock or a tussock with worms curling in their bills for longer than most people care to wait. They are double-brooded and their song goes on undiminished into late July and fragmentarily into August even after the flocks are beginning to form. Every summer I am grateful that the pipits' nesting-time is so prolonged, for some of the great mountain grass-slopes would be very quiet and lifeless in July without the pipits and their families flitting and calling everywhere and many still courting and in full song. Some of these pipit songs are very fine. Perhaps they sound a bit thin in the lowlands in May if you hear them against a chorus of thrushes, blackbirds, tree pipits, robins and warblers. But up in the mountains, especially on a fine summer evening when all is still and the hills are cut out black against a yellowing sky, then the falling songs of the pipits seem particularly beautiful. It has often seemed to me that the song and the accompanying flight, finishing in a long glide over the grass, is more prolonged and richer in the evening than at any other time.

From June onwards young pipits are everywhere. As they become independent they spread all over the hills until no matter where you walk they fly up before you. They develop the curious habit at this season of taking long siestas in little hollows they shape in the grass. In these bowers they evidently crouch for long periods, judging by the quantity of droppings which accumulate there. Many is the time, seeing a pipit burst from the centre of a tuft of grass, that I have thought to have discovered a nest only to find one of these siesta couches, which are of course very crude compared with the true nest, for a meadow pipit's nest is perfectly cupped and nearly always lined with very delicate, light-coloured straws, against which the chocolate-coloured eggs make an attractive contrast.

In a summer's walk across the hills it is usual to chance upon one or two pipits' nests by nearly treading on the sitting bird. In this way I have found as many as six in a few hours yet at other times, deliberately searching for nests, I have failed completely. Only once have I found a meadow pipit's nest by spotting the bird in the nest. I was bending down to examine a moss on a rock and saw her large dark eye first before I connected it in my mind with a bird. It is rare that one gets within two feet of the nervous pipit but this one was brooding callow young, which is when birds sit tightest. I was struck by the very great beauty at close range of a bird that looks so sombre at a distance. The delicate pencilling on her crown, the suggestion of an eye-stripe, the bolder streaking down her back, the steady shining eye, the thin, sensitive beak: these went well with the soft green sphagnum moss surrounding the nest and the silver and purple lichens on the rock beside. Much of the beauty of the mountain birds and their surroundings is made up of such quiet colours.

Most abundant of the upland birds, the meadow pipits

inhabit not only grassland but also heather, bracken, rushes and the wettest bogs. But they are abundant only from spring to early autumn. By winter only a few are left which somehow manage to survive on the high ground provided the weather is not abnormally severe. What an insect-eater such as a pipit lives on in the open mountains at midwinter I cannot imagine, unless it turns seed-eater for a while. By mid-March the returning pipits begin to approach the uplands but at first remain in flocks—sometimes a hundred or two together—round the lower skirts of the moors, not moving up until the end of March or early in April when the temperature rises to springlike levels. At that season the flocks are very restless and excitable and there is a lot of calling and chasing. Once in the hills the flocks quickly break up and do not re-assemble until late summer.

Cuckoos, frequenting many different types of lowland and upland country, are perhaps commonest in Snowdonia on those rough, brackeny, thorn-scattered hillsides that occupy so much of the Park between 600 and 1,500 feet. I have occasionally heard a cuckoo up moorland valleys at nearly 2,000 feet but never met with one on the high open tops. Up there, although they would find enough pipits' nests to lay their eggs in, cuckoos would presumably find insufficient food, for they are voracious feeders, as I saw one day when I managed to get close to a female cuckoo. She was perching clumsily, as cuckoos do, on old bracken stems, balancing with the aid of now one wing, now the other. Her red eye was intent on the ground below and about every ninety seconds she made a quick dive to the grass, seized a caterpillar and flew back to the bracken, moving on about a yard each time. The caterpillars were all the same: the large, hairy, slate-grey caterpillars of the drinker moth. Each one was taken by the end, never across the middle, and was held dangling from the cuckoo's beak a few seconds, then vigorously shaken and immediately swallowed with one quick backward jerk of the head. I watched the cuckoo take a score of these big caterpillars from a few square yards of hillside and

I do not know how many she had eaten before I got there. Her keenness of sight was wonderful, for most of these caterpillars were pulled out from deep in the grass.

The male cuckoo arrives here with fair regularity within a few days of April 21. He is the one migrant whose voice all country people here notice and comment on and almost invariably they speak of him as 'she', for both the Welsh words for cuckoo, *cog* and *cwcw*, are feminine nouns. These first cuckoos do not usually stay but pass on elsewhere and a week or more may go by before we hear the next arrivals. Then one day we go to the hills and find cuckoos all round us. There may be only three or four of them, but so persistently do they call when first settling in that it takes but a handful of cuckoos to make the land seem swarming with them. Through May and June they are constantly about their chosen hillsides, hurrying along the same short beats, chasing, calling, feeding, or being chased and chivvied by wheatears and pipits. Then, quite abruptly, it is all over. There comes a day in late June (in a late season it is not until mid-July) when, crossing the hills, you hear no cuckoo at all, not a male calling or growling nor a female bubbling. The last egg has been laid and the parents have gone away Africa-wards till spring shall come again. But that is not the end of the story. For the young cuckoos are left to be reared by their pipit foster-parents and sometimes you can find them by hearing their wheezing cries for food as they lie in their nests in the grass or later as they perch conspicuously on rock or gate, imploring any passing bird to feed them. Then, once independent, the young cuckoos do not linger in the hills. Urged by instincts of which we understand nothing, guided by means about which we can only make guesses, they depart, each young bird alone, on a three-thousand mile journey south.

Though there has been a serious decline in red grouse since the beginning of the century they are still widespread wherever there are stretches of heather-moor above about 1,200 feet. Grouse are most numerous on the Berwyn section of the Park at about 1,500 feet. After that the best grouse areas are probably the moors east of Ffestiniog and on the Rhinogs over in the west. There are some keepered moors, especially on the Berwyn, but far fewer than before the First World War. The scattering of now ruined shooting-boxes still to be seen in the heather country is a reminder of pre-motoring days when to go shooting on the moors was quite a safari and food and shelter were a necessary provision there. During this century not only have the grouse declined but also many of the larger estates. The result is that gamekeeping has also declined and the keeper's gun, poison and poletrap are things of the past on most of the moors. But not on all. Grouse are splendid birds to see and hear among the heather; but it is more difficult to work up enthusiasm for the grouse of keepered moors where you know that, in defiance of the law, the keepers (usually with the full knowledge of their employers) are ready to destroy any buzzard, kite, short-eared owl or harrier that may chance their way. The hen-harrier in particular is a bird which, given all possible protection, might well establish itself as a breeding species on the Snowdonia moors. It is perhaps a reflection of the recent increase in the number of hen-harriers in Scotland that a pair has nested in Snowdonia at least since 1962. Let us hope that this is the beginning of a serious attempt at colonisation, or rather of a return, for this harrier used to breed in Snowdonia a century ago but was undoubtedly wiped out by grouse-keepers. A curious but well authenticated breeding record is that of the pair of Montagu's harriers which nested on the Merioneth moors in 1900 for this is not usually an upland nesting species. This pair bred on heather-moor 4 miles south of Bala. The eggs were taken and both parents killed. The height above sea-level of this nest is not stated in the record but on internal evidence it appears to have been at about 1,500 feet. There is only one recent record of Montagu's harrier breeding in the Park. This was near the coast in 1951 when three young were reared.

While the grouse has been declining another conspicuous moorland bird has been increasing. As a breeding species the black-headed gull was evidently at a low ebb during the later nineteenth century (we have no reliable evidence earlier than that), but it increased in the first quarter of this century and has probably remained fairly steady in the Park since the 1930's. Except in one coastal colony (the largest in the area), this small gull breeds in a scattering of minor moorland communities which usually occupy a site for several years running, then change quite whimsically, or so it seems, to somewhere else. The largest colonies tend to be the most stable, provided they are little disturbed. Here and there pairs nesting alone are not unusual, especially as outliers from nearby gulleries. The breeding places are usually swampy-edged lakes or morasses in quiet parts of the hills below about 1,800 feet. Most of the lake-edge nests are placed on tussocks surrounded by water or quaking bog, some even floating anchored among rushes or horsetails. But on islands the nests are usually on quite firm ground. Clearly the main attraction of these upland lakes is that they provide nesting sites fairly well protected from foxes. That there is very little to eat in some of these peaty pools and bogs is no objection from the gulls' point of view because they do not seem to mind how far they have to go for food. You can see them all through the breeding season flying over the hills or the lowland fields very far from any nesting places. They evidently feed their young on a wide variety of animal and vegetable food, for you can see them foraging in many different habitats.

Black-headed gull colonies are very noisy, attracting attention in a way that would seem to nullify the care with which they choose the remotest sites. But foxes would detect them by scent even if the gulls were silent and it is probable that even a few gulls, by keeping up as much noise as possible, can give the impression that there is a multitude of them and so deter some of their enemies. Whatever their use, the wild harsh cries of these gulls are very beautiful to hear distantly across the moors. And the gulls themselves are a splendid sight exploding in a white cloud from their nests as you get near them. But it must be remembered that in these days of cars and good mountain roads most of these gulleries are no longer the remote places they have always been in the past. They are very open to human disturbance and therefore people who visit them should make their stay as short as possible for their intrusion may be only one of several the same day.

Very few are the pairs of golden plover breeding in Snowdonia when you think of all the miles of moors and hills available to them. And these few are so shy and elusive and so well camouflaged among the tussocks of grass and heather that it is not surprising that even regular mountain walkers rarely seem to notice them. If noticed at all, they are usually only a voice in the moorland wilderness, a distant, soft, sad-sounding *peee* or sometimes *tlo-eee*. This is the sentinel note you hear as you cross their territories any time from when they arrive in March until they depart in late summer. You may search quite a time before you see the bird, perhaps standing very still on a mound and watching you carefully. Then, if you look at him through binoculars, you will be surprised, if you know golden plovers only as greyish winter birds of the coast, to see the black belly-patch and the rich checking of yellow and black all over the wings of this plover in breeding dress. If you try to approach him, as he hopes you will because he has a mate on eggs not far off in another direction, he will by walking and short low flights gradually lead you away, mysteriously disappearing at last and leaving you not at all sure where it was you first saw him.

When the golden plovers return to their breeding grounds in early spring they are often in small flocks. So in April you may think you have found what looks like a promising breeding colony. But go there a month later and you will be lucky to find even one pair there, the others having spread out over a vast reach of surrounding country and you may have to circle round for miles before you hear the songs of other golden plovers and so get some idea of the extent of their territories. The song is a rhythmic *too-roo, too-roo*, varied by a rarer but delightful bubbling medley of quickly rising and falling notes. The voice of the golden plover is a frequent sound on their territories in April, but by June I have known silent days when I have not heard a sound from a golden plover till after dark.

The golden plovers that occur here and there in small flocks on the hills in winter are almost certainly of the race that breeds in the high north of Europe, for the British breeding stock is reckoned to migrate south in autumn. Occasionally individuals of the northern race—they are very distinct with their bold, constrasty markings—get mixed up with parties of the returning southern race in spring and stay to breed.

For me one of the most exciting birds to find nesting on the moors of Snowdonia, not only because it is a delightful bird but also because it is a rarity here, is the dunlin. To find dunlin nesting is to meet the southern fringe of the arctic. It is as good as finding the purple saxifrage. For the dunlin that breed as far south as Wales are on the tip of a tongue that stretches out unexpectedly from the far northern latitudes that are the breeding grounds of nearly all the world's great population of this little wader. And often when the dunlin arrives at the end of winter there are parts of these moors that have very much the look of the arctic tundra—miles of treeless country patched with melting snow and scattered with ice-edged lakes and peaty pools. It is little wonder that waders predominate in these regions of Snowdonia and that seven species of them breed there.

The dunlin adds to his attractiveness by being by far the most approachable wader breeding in the Park. Once when I was sitting quietly at the rim of a stony-edged lake a dunlin walked towards me round the water's edge feeding hurriedly, head down in the accepted dunlin manner. So preoccupied was he that he hopped right over my foot and went on round the lake-edge without noticing me at all. That of course was exceptional behaviour but often the newly arrived dunlin are very confiding. They develop more caution as the season advances and when they have eggs or young I have had them flying anxiously round me or perching on hummocks uttering a vibrant *zweer-zweer-zweer*. The dunlin's most delightful note is the long, falling, purring trill you hear from them (both sexes use it) when they are gliding over their nesting territory.

Breeding dunlin are not only extremely thin on the ground in the Park, they are also very oddly distributed: they are apparently confined to Merioneth, for as far as I know there are no Caernarvonshire records. But there are so many spots in Caernarvonshire that look suitable for the dunlin that it is hard to believe there is in truth none there. So there is an obvious challenge for energetic bird watchers to take up. Its scarcity obviously puts the dunlin high in the list of those of the Park's breeding birds that need special protection. It is not a question of protecting the eggs from collectors—this would be almost impossible on remote moorland—but of conserving their breeding habitats. Drainage or afforestation of peat bogs near lakes: this is the greatest threat to the dunlin. And nature conservationists need to be alive to these dangers so that steps can be taken before irretrievable damage is done. For this reason anyone finding dunlin breeding in the Park would do well to report them to the Naturalists' Trust concerned (see appendix).

Along with golden plover and dunlin there is one other high-level breeding wader in Snowdonia: the snipe. By high-level in this connection I mean upwards of approximately 1,600 feet. Some snipe nest up to at least 1,800 feet on remote moors, where their weird bleating as they dive unseen in the darkness is the only sound in the stillness of the night. But of course it is a frequent daytime sound as well. It is a sound I love to hear when camping out on the moors; and incidentally it is in dusk or darkness that many of these upland wader species are most vocal and therefore most easily detected. In the full glare of the afternoon sun they can all be very skulking and difficult to find. Snipe are not of course confined to moorlands. They breed in bogs and marshes at all levels down to the coast and some pairs have a curious preference for nesting close to villages and farms. I have several times seen a snipe's nest within thirty or forty yards of a cottage gate where there was very often the noise of children, dogs and traffic. Yet in the neighbourhood were several quieter bogs that had no snipe in them. Besides the bleating noise, which is feather-produced, there is the snipe's call-note, which can be one of the most persistent sounds in all nature. I once heard a snipe calling *chippa-chippa-chippa-chippa* with monotonous rhythm for hours on end during a day and half a night. It was one of those days of warm, thick mist so common on the hills in summer.

The upland farmers and shepherds may know little about golden plovers and still less about dunlin. But all know the curlew and are glad on that day in late February or early March when they hear its first plainings and bubblings over their fields. For the next few weeks the lower hills, especially where there are bogs or marshes, will be lively with the curlews' music, deep wailings that rise and quicken to ecstatic crescendoes followed by falling, dying notes which may rise again into a clear, triumphant *courlee-courlee-courlee* many times repeated. Not that curlews have any season quite without song. In the depths of winter fragmentary bubblings from the coastal flocks give us sharp memories of the hills in spring. The fullest courtship song lasts a few weeks only. Already in late April it begins to get less as egg-time arrives and during the month of incubation a pair of curlews can be very quiet and unobtrusive. Then the eggs hatch and alarm notes begin to multiply: you can always tell when curlews have young by the agonised yelpings with which they mark your approach or dive at passing crows, kestrels and buzzards.

The young curlews, before their feathers cover them, are delightfully woolly and mottled and have straight beaks. They scurry between the tall rushes on long, pale-blue shanks. Gradually they are led away from the nesting area by the parents and I have known a family a week after hatching to be over half a mile from the nest. By mid-June many young curlews can fly and the families are joining up with one another. By the second week of July you may see flocks of thirty or forty. By then the adults are in moult and leave their beautifully patterned feathers round the margins of lakes. For each flock has its favoured water where the birds stand, sleep, walk, preen or scratch for undisturbed hours of the long days of summer. These moulting flocks can, for curlews, be unbelievably quiet. Almost entirely gone now are the yodellings and the switch-back flights of spring. Yet watching them as they drowsed in the summer's heat, I have seen a mysterious air of excitement tremble through a resting flock and give rise to a burst of bubbling cries that has quickly subsided. September sees the departure of most curlews from the high ground. By winter practically all are on the coast or gone off on far migrations. The curlew breeds abundantly in the Park on many types of rough ground, but also in cornfields, from sea-level to about 1,200 feet. There are far fewer between 1,200 and 1,600 feet, and those seen higher than that are probably feeding, not nesting. There is evidence that more are now breeding on lower ground and fewer in the hills than earlier this century.

There are few lakes of any size with firm shores that are not enlivened from late April to August by the cheering notes and attractive flight of the common sandpiper, a summer wader that is also widespread along moorland streams and lowland rivers right down to the coast. Its distribution tends to be patchy, however, because the pairs are sociable with other pairs and form little communities—if community is the correct word for a type of association in which each pair strictly maintains its private territory. The existence of these communities is most striking along a river where you may walk miles without seeing a sandpiper and then in a couple of thousand yards pass through the territories of several pairs.

There is certainly no missing the sandpipers once they have arrived for they advertise themselves every few minutes by their excited call-note *peep-ip-peep* or by the rapidly repeated song-phrase *kirkitly-wee-wit* as they run about the stones chasing and displaying. All waders have attractive shapes and this one particularly. Compared with, say, a dunlin, a sandpiper looks more slender, more upright and longer in the shank. He seems much livelier because of his ever-bobbing stern and fluid movements. He is always good to see, whether he stands on a bridge-rail shouting defiance from six feet away; or feeds briskly in the shallows flicking out pebbles with sideways jerks of his bill; or, having found some food, dips it carefully into the water to wash it. The nest is either hidden under a tussock or, less often, simply lies open to the sky like a ringed plover's. Sometimes a pair uses the same nest in successive seasons. To me the eggs are more delicately beautiful than those of any other bird: slightly larger than blackbirds' eggs, they are usually warm buff heavily blotched with a reddish chocolate-colour with pale filmy under-markings of bluish brown. The downy chicks too are most beautiful, with freckled dark-grey backs and whitish undersides and ludicrously large silver-blue legs and feet. Even though only just hatched, they obey the *peep* alarm note of their parents immediately and disappear miraculously as soon as they hear it.

The only other regular breeding wader of the hills, the peewit (or lapwing) is now far less a moorland bird than in the first quarter of this century. In those days, all accounts agree, peewits nested everywhere in the wetter parts of the moors and their song-flights and calls were an essential part of springtime in the uplands at least as high as 1,600 feet, while quite a few nested up to 2,000 feet. But when numbers declined steeply in many parts of Britain, these populations at the topmost fringe of the peewit's range drained rapidly off the hills. Today very few breed as high as 1,600 feet. There are more from 1,000 feet down to sea-level but they are sparse almost everywhere.

Three last waders remain to be mentioned as possibilities on the hills in spring or summer: the redshank, the oyster-catcher and the dotterel. The redshank is very much a coastal breeding species in Snowdonia but here and there it goes to the moors to breed. It seems reasonable to hope that the oystercatcher, since it is breeding increasingly inland in Scotland, will establish itself inland in the Park in the near future. Possible signs of this happening are that oystercatchers did bring off a brood or two a few years ago at Trawsfynydd Reservoir and have been reported in the nesting season from the middle reaches of the Dovey. They should certainly be looked out for. The rare dotterel has occasionally been recorded on Welsh mountain tops on the spring and autumn migrations. It is a bird I have long hoped to meet with, for there could be few more exciting species to see on some remote Welsh summit.

Only two kinds of duck, the mallard and the teal, breed regularly in the hills. Both prefer lakes with plenty of vegetation and teal especially seem happiest in water almost entirely choked with horsetails or bottle sedge. But as so many of the mountain lakes are deep, stony-edged and almost weedless, the breeding waterfowl are very few for so large and well watered an area as Snowdonia. Whatever duck you find on these lakes are always difficult to approach because of the lack of cover. And they are very wary. The mallard drake is particularly alert and long-sighted. Try to get near him and his long green neck soon shoots up, his head stiffly held, beak forward and level, though his mate may remain at ease, perhaps upending for food. For maybe two minutes he watches you rigidly. Then he quacks just once, softly. Instantly his mate is as alert as he. You keep still; so do they, weighing the danger. Then they rise with a sharp splash into flight against the wind, the duck often half a yard in front. In a few seconds they are beyond the end of the lake and now they circle and come racing back downwind on whistling wings, the drake now leading. Three or four times they fly round, spiralling higher, getting their direction. Then away they go across the moor to some lake they can see beyond the ridge. They show such lack of attachment to the place in the middle of the nesting season that they are clearly a non-breeding pair. I give these details because this is so typical an experience with mallard in the hills in summer. There must be a large proportion of them not breeding every year. The mallard that are breeding behave quite differently, showing reluctance to go very far. And of course, when their young are out the ducks are extremely demonstrative, performing the broken-wing trick and other diversionary tactics.

The teal are usually in little communities of two or three pairs. When approached, they rise in a flock, springing very neatly into the air and racing down the lake, their green wings flashing as they turn. Like the mallard, they too circle and head away across the hills. But not far. In a few hundred yards with a sudden change of direction they all plunge steeply to a bog pool or a rushy stream and disappear. Then if you follow them and put them up again they will probably fly straight back to the lake. Their nests are particularly hard to find because, while a mallard will often lay openly, right at the water's edge, a teal may go half a mile away into the heather and hide her eggs in thick cover. I have once found eggs in mid-April but most teal eggs are perhaps laid in early May, for I have most often seen ducklings in early June. But some are even a month later than that.

As some of the weedier, shallower moorland waters suit mallard and teal, there is always a hope of finding wigeon breeding too, for though it usually nests much further north the wigeon is well known for its propensity for breeding sporadically in places well outside its normal range; and in recent years a pair has bred in North Wales, only a few miles outside the Park. Only two other waterfowl nest regularly in the hills: moorhen and little grebe. Neither is common but these upland moorhens may be a little more numerous than appears for they can be extremely secretive. So up to a point can the grebes, but at times they have a loud trilling chorus that advertises their presence. A few pairs of great crested grebes bred for years on Trawsfynydd Reservoir before the atomic power-station was built—the only breeding place of this grebe in Snowdonia. In winter there is a scattering of surface and diving duck on the lakes and reservoirs, also occasional grebes, divers and wild swans. But owing to the acidity of many of these waters they contain insufficient food to attract many waterfowl.

Two wagtails, the pied and the grey, are widespread in the hills from spring to autumn. The grey, restricted to the vicinity of streams and rivers, is much the less numerous, for the pied wagtail nests not only along streams but on rocks away from water. The pied is also a great frequenter of upland farms and so is most abundant from about 1,000 feet downwards. The grey wagtail also occasionally forages in farm-yards that are not far from water and sometimes even nests in the walls of farm-buildings. Most wagtails come down off the high ground in autumn to spend the winter either in the nearby lowlands or far away to the south. But in a mild winter you may meet individual wagtails in the hills, the pied occasionally, the grey more rarely. Both sing sporadically during winter.

There are few more gladdening sights on moorland streams at the end of winter than the year's first grey wagtail, for this is a species of outstanding beauty, the males especially, with their slate-grey backs and brilliant yellow underparts, their very slender bodies and long tails and their irrepressible liveliness. People understandably persist in calling them yellow wagtails, but the true yellow wagtail is a rarity in Snowdonia and highly improbable on bouldery, rushing hill streams. Grey wagtails constantly shoot up and down the stream giving their sharp flight-note, or they stand on mid-water rocks babbling out their songs in harmony with the hurried water of the stream. The males are not always in full breeding dress when they arrive. One I watched displaying before his mate had a pure white throat as late as March 18. By March 26, when the hen was building, the male was showing a little black on the throat. Not till April 14 was his throat entirely black, contrasting beautifully with the yellow breast and divided from the grey cheeks by a bold white stripe. March is very early for grey wagtails to be nesting and on the high ground you are not likely to see them carrying straws before late April. The nest, often more deeply hidden than the average pied wagtail's, is very much like it except that white hair is the almost invariable lining. This fixation on white hair is so strong in the mind of the hen grey wagtail that she will go far to seek it.

It becomes a habit to speak of dippers and grey wagtails in the same breath, so often do they nest near each other. But two birds of the same habitat could hardly be more unalike in appearance or habits. Watch them building, for instance. Only the female wagtail builds, delicately selecting fine materials from the streamside and quietly slipping off with them to her nest. It is quite otherwise with dippers. They go at their nest-building with gusto, both of them hurrying in with beakfuls of stuff, and while the wagtail shapes her neat little cup, the dippers throw together what looks like an outsize in wrens' nests. One pair I watched building would sometimes, instead of flying off downstream, simply drop on to the water and let the current carry them down to a mossy rock in mid-stream. They attacked this moss with energy, straining backwards to tear it off, often skidding or overbalancing into the water and swimming round looking really weighed down with their moss loads, yet still scrambling out for more. That such normal-looking land birds should have taken to water so thoroughly is remarkable; they not only swim on it but dive through it and walk under it. Dippers breed in the Park from near sea-level to the high mountain streams. They usually begin breeding in April but much earlier nests are on record. H. E. Forrest (1907) described the dipper as 'common on suitable streams throughout North Wales'. But I find the dipper widespread but rather thinly distributed, very much less common, for instance, than the grey wagtail. Possibly there has been a decrease this century.

The sight of rooks streaming up to the hills on mornings of early June is an annual sign that spring is nearing its end. A few adult rooks forage on the hills in April and May but it is about the first week of June that leisurely, cawing flocks of parents and young begin to make daily journeys up from rookeries miles away down the valleys. It is strange, yet quite pleasant, suddenly to find the quiet uplands invaded by these flocks of large black birds scattered untidily across whole mountainsides, cawing out their eternal conversations with every variety of note from deep throatings to high-pitched squeaks. A few weeks later the chorus increases when the jackdaw families are out of the nest, for they too join the twice-daily processions up to the hills in the morning and back to the valleys in the evening. Families of carrion crows, too, sometimes add themselves to the crowd. Clearly there must be great reserves of insect-food in the high pastures to support all these invaders, for they seem to feed incessantly, the young following their parents closely and running to them every few minutes with gaping beaks, flapping wings and hunger cries.

Quite a few other birds, free to wander after the nesting season, go to enjoy the brief summer harvest of the summer uplands when insects are multiplying, grasses are seeding and bilberries and then rowans are fruiting. But none of these species—woodpigeons, mistle thrushes, redstarts, greenfinches, redpolls and others—go to the hills with anything like the regularity and in such numbers as the rooks. In August there is a trickle of small migrants through some of the passes, especially willow warblers and whitethroats, flitting alongside mountain tracks from one clump of heather to another. But it is not a well marked movement and is not seen at all some years.

August sees a remarkable move-up of herring gulls from the coast, a movement distinct from the foraging of gulls on Snowdon. In great straggling flocks they make daily journeys from the coast to the hills. They often go much higher than the crow hordes of early summer. Nor is it food that these gulls seek, for often they settle on arid slopes or bare mountain tops. I have spent the whole of an August day on a stony flat-topped ridge at nearly 2,500 feet with a flock of over a thousand herring gulls. They ate very little but simply stood in the sun or walked slowly about. They were heavily in moult, silent and listless. Why they had gone up there to moult, I cannot suggest. But they were beautiful to see as they dispersed seawards towards evening, drifting in twos and threes off the sunlit top down towards the deepening shadows of the valleys.

In the past few seasons the herring gull—just a pair

here and there—has begun breeding inland in the Park, at the margins of a few of the upland lakes. As this is a gull that has increased enormously elsewhere in Britain in recent years, are we now to see a general invasion of the hills? Inland breeding by either of the black-headed gulls is also a possibility. Great black-backed gulls used to breed at a few lakes in the eighteenth and nineteenth centuries. Today they are fairly frequent about the lower sheepwalks, especially after a severe winter when there is carrion mutton lying about, for they are often vulture-like in their feeding. A few apparently mature pairs of great black-backs frequent upland waters all the season but do not breed. There is a marked migration of lesser black-backs over the hills especially in spring, usually in pairs or very small flocks. They never seem to be in a hurry, for their flight is leisurely and they often come down to feed or to rest.

Single cormorants or sometimes small parties of them can frequently be met with at lakes at any height provided there is fish in them. If you disturb cormorants they usually merely fly over to the next lake; but once I saw two rise from the water and spiral round higher and higher; they were almost beyond naked sight when they finally made off. Cormorants are so much more successful than anglers at catching fish that it is scarcely surprising that angling associations persecute them. But to do so is probably very short-sighted policy. Angling associations should never forget the experience of their colleagues in the United States who, having destroyed all the mergansers on a lake, found the fishing soon deteriorated but that it improved again after the mergansers were given protection. The reason was simply that the mergansers fed entirely on young fish and in so doing removed just the right quantity of them to allow the remainder to grow up into a healthy population of well-developed adults.

By late November the high ground is becoming cold, bleak and wintry and only the few true winter birds are left on the hills: grouse, raven, crow, buzzard, dipper, wren, redwing, fieldfare (both these northern thrushes feed on open moorland, especially wet places, as long as the weather is not severe), a few meadow pipits, a scattering of snipe, golden plover and curlew and here and there a short-eared owl, a passing chough, a few white-fronted geese or perhaps other more casual species. One winter's afternoon up Talyllyn Pass I saw a great grey shrike singing a scrapy song perched on a roadside fence-post at sunset. This is a rare bird in Wales, but if you are lucky enough to see one there is a distinct chance of finding it in the same place again for this shrike sticks to its winter territory once it has adopted one. A great grey shrike I once had under observation kept to the same few acres of field and thorn-scrub for at least a month. Another rare winter casual is the snow bunting, which is more likely on the coast but occasionally seen about the hills. I once saw about eighty in a beautiful, dancing, tinkling flock hurrying past me down a valley one winter afternoon; but such a large number is quite exceptional.

In a severe winter the high ground can become so snowbound or scorched by the east wind as to be quite uninhabitable for practically all birds. Then even the red grouse—a whole moor of them—may shift a few miles to somewhere where they can find living heather. Occasionally, after such a move, they have been known never to return, having settled happily in their new quarters. In that sort of winter many of even the most hardy mountain birds come to grief. For when they have quitted the hills they find conditions no better in the valleys or even on the coast. Sometimes the estuaries, the saltings, even the sea-shore is frozen hard. Then many birds die. This happened, according to Pennant, in January 1776 when the shore was 'for miles together covered with dead birds'. It has doubtless happened in many other winters. In Snowdonia we saw something like it in 1947 and 1963. After such disasters there follow several seasons when many species are reduced in numbers and some are quite absent locally. Eventually they all seem to recover but can we be quite sure that these disasters do not have long-term effects on the status of some birds? What is certain is that the numbers of some vulnerable species hang on a complexity of delicate threads. For thousands of years they may have only just survived the severest winters and other natural hazards and are now perhaps threatened by some disastrous alteration of their habitat or unintentional poisoning of their food by man. It is clearly imperative for us to be always on the look-out for such threats to wild birds.

CHAPTER SIX

Nature Reserves and Conservation

The Snowdonia National Park, with its wide variety of rocks, soils, heights, aspects, climate and consequently plants and animals, is a region of particular value for ecological studies and nature conservation. Its former inaccessibility and remoteness from large centres of human population long protected it from severe exploitation and disturbance. Now those times are gone and many interests clash in their attempts to use parts of Snowdonia for their different ends. It was high time for such a move when the Nature Conservancy was established in 1949 to safeguard what it could of Britain's remaining haunts of wild life. Its first nature reserve in Snowdonia was at Cwm Idwal (1954). Since then fourteen others have been added and, it is hoped, many more will be created. A list and an account of the reserves already in existence is given below. It will be seen that there are five mountain reserves and eight woodland reserves; the other two occupy shore and sand-dune areas. Application for permits to visit the reserves should be made to: The Nature Conservancy, Penrhos Road, Bangor, Caernarvonshire.

NAME	NATURE	ACRES	LOCATION	IS PERMIT REQUIRED?
Coed Gorswen	woodland	33	4 m. S. of Conway	Yes } away from public footpath
Coed Dolgarrog	woodland	170	7 m. S. of Conway	Yes }
Snowdon	mountain	eventually about 5,000	11 m. S. of Bangor	No
Cwm Idwal	mountain	984	5 m. W. of Capel Curig	No

Cwm Glas Crafnant	mountain and woodland	38	1½ m. N.E. of Capel Curig	For enclosed woodland
Coed Cymerau	woodland	65	1 m. N.W. of Ffestiniog	Yes, away from public footpath
Coedydd Maentwrog	woodland	169	½ m. N. of Maentwrog	No
Coed Camlyn	woodland	57	½ m. S. of Maentwrog	Yes
Coed Tremadoc	woodland, cliff & block scree	49	1½ m. N.E. of Portmadoc	Yes. Special permit for rock-climbing
Coed y Rhygen	woodland	52	W. side of Trawsfynydd Reservoir	Yes
Morfa Harlech	shore and dunes	1,214	2 m. N.W. of Harlech	Yes, above HWMOT*
Rhinog	mountain	991	5 m. E. of Harlech	No
Morfa Dyffryn	shore and dunes	500	4 m. S.S.W. of Harlech	Yes, away from public footpath and above HWMOT*
Coed Ganllwyd	woodland	61	5 m. N. of Dolgellau	No
Cader Idris	mountain and woodland	969	4 m. S. of Dolgellau	For enclosed woodland

*High Water Mark of Ordinary Tides

This chapter outlines the main features of the reserves, some of which are dealt with more fully in the topographical section of the book. The reserves have been surveyed in detail by the Nature Conservancy from whom further information is available.

The Mountain Reserves

In the five mountain nature reserves—Snowdon, Cader Idris, Cwm Glas Crafnant, Cwm Idwal and Rhinog—

you can find examples of practically the whole range of the Park's upland rocks, soils, flora and fauna.

Snowdon. The Snowdon nature reserve is of outstanding importance as the classic locality for Ordovician vulcanicity. It is also a site of rare mountain plants which ever since botanical studies began in the seventeenth century have been so keenly sought after by collectors that today rare species such as holly fern, the woodsia ferns, Snowdon lily and others are much harder to find than formerly. As this book goes to press, only south and eastern Snowdon are reserved but it is hoped that soon almost the whole mountain will be included in one reserve of some 5,000 acres. It will then include all the richest botanical sites: the cliffs of Clogwyn Du'r Arddu, Cwm Glas and Cwm Dyli. Access to Snowdon by the general public is virtually unaffected by its nature-reserve status but anyone wishing to collect plants now needs a permit. The presence of full-time wardens should put an end to the long despoliation of Snowdon's rare wild flowers and ferns. More information about Snowdon will be found especially in chapters 2, 3 and 7.

Cwm Idwal. This was naturally the first nature reserve to be declared in Wales. It is not only the supreme site for the study of Welsh mountain vegetation, it is also holy ground for the history of early botany and geology, and is therefore the best documented of all the reserves. Ray, Lhuyd, Pennant, Sedgwick, Darwin, Ramsay, all came to Idwal for their researches. So did many of the pioneer travellers who looked with awe at the lake, the boulders, the hanging gardens and, above all, at the chasm of the Devil's Kitchen. Incidentally, it was here in the 1890's that the pioneer alpine climbers came to practise their art on the Devil's Kitchen cliffs, men like J. M. Archer Thompson, Harold Hughes, Owen Glynne Jones and the Abraham brothers.

Cwm Idwal reserve embraces not only the cwm itself but also the higher slopes above to the tops of Y Garn and Glyder Fawr and so includes two more lakes—Llyn

Clyd and Llyn y Cwn, and two more crags—Clogwyn Du and the Gribin. As on Cader Idris, there is an assortment of acid and basic rocks with a consequent variety of plants. But the glory of Cwm Idwal is the celebrated Trigyfylchau which is riven from top to bottom by Twll Du, the Black Chasm (Devil's Kitchen), on either side of which the downfolding of the rocks (the Idwal syncline), comparable with the downfolded rocks of the top of Snowdon, is so clear to see. It is on these rocks of what geologists call the 'bedded pyroclastic series' (andesitic basalt and calcareous pumice tuffs) that the main wealth of the flora is concentrated. Here, as well as on other Caernarvonshire cliffs, Lhuyd found quantities of the tiny lily later called Lloydia in his honour; and here it still grows, together with a wide range of arctic-alpines (see appendix) and those more typically lowland species which often flourish on the richer mountain ledges: hairy rock-cress, ox-eye daisy, burnet saxifrage, meadowsweet, globe flower and many others. Here William Bingley (1798) was surprised to see 'exceedingly luxuriant' sheets of another lowland plant, the moschatel. But the moschatel is a surprising plant: it not only consorts with alpines on many of the world's great mountain ranges, it thrives in the Arctic as well. You can see another familiar lowlander in Cwm Idwal: the coltsfoot, which is not infrequent at the edges of mountain springs on base-rich soils in Snowdonia (it is on Rhobell Fawr, Cader Idris, Dduallt and many other places). It looks as if there are two races of coltsfoot—a mountain race and one that is characteristic of railway banks and waste places.

In remote prehistoric times the flora of Idwal was unquestionably very different from what it is today, but it was still rich in calcicoles. It is strange to think that their pollen still lies in the cwm, retained there by the remarkable preserving action of peat. Among the plants that have been identified in Cwm Idwal by pollen-analysis are rock-rose, hoary plantain, salad burnet and Jacob's ladder, all distinct calcicoles. Presumably they throve there during the Atlantic period (about 6000 B.C.) when the climate was warmer than now.

As for animals and birds in Cwm Idwal, there are 'wild'

goats, foxes and stoats. There are lizards and palmate newts. In the cliffs, ravens and ring ouzels; on the slopes, wheatears and meadow pipits; round the lake, sandpipers and grey wagtails; in the lake, trout and minnows; and on the lake, from autumn to spring, occasional small parties of pochard, goldeneye, tufted duck and whooper swans.

Cader Idris. The best mountain nature reserves are often in the most spectacular and beautiful places. It is so on Snowdon and in Cwm Idwal and also in the Cader Idris reserve which occupies Cwm Cau, the splendid corrie gouged with such perfection out of the southern flank of the mountain, a corrie with great cliffs encircling a deep lake from which a river cascades from pool to pool for several hundred feet down through steep, delightful woodlands. Cwm Cau is a show-piece for the geographer, the geologist and the ecologist. It is the complete hanging valley full of the evidences of glaciation. Its cliffs are important for the study of the volcanic rocks as well as mudstones; and where they are calcareous and have water dripping down, as on the south-west and the north-east of the lake, there you can see an interesting plant community including green spleenwort, mountain sorrel, rose-root, lesser meadow-rue, alpine scurvy-grass, moschatel, aspen, Welsh poppy, mossy and starry saxifrages and other good plants. The ecology of Cwm Cau is best considered in relation to that of the corries on the north side of the mountain and it is much to be hoped that they too will eventually become part of this reserve as was originally intended. An important point to keep in mind in such ecological studies is that here and there one is surprised to find calcicole plants on what may have been mapped as acid rocks (as on the upper acid rocks of Geu Graig); or calcifuge plants on what are lime-rich rocks.

So the botanist must guard against being guided solely by geological maps. He must leave it to the plants to tell the truth about their own ecology. A lime-rich water, for instance, might seep down over acid rocks and provide a quite unexpected habitat for calcicole plants. Another interesting aspect of Cwm Cau is that it comes right at the bottom of the figure 5, the shape in which Snowdonia's volcanic rocks lie on the map (see page 44). Hence this corrie is the most southerly place in the Park where calcicole plants are in any quantity. South of Snowdonia stretch many miles of upland country where calcicoles are rare.

Cwm Glas Crafnant. At the head of Crafnant at the east end of the Carneddau is the spectacular valley of Cwm Glas, which, like so many high corries, looks particularly fine when the jagged rocks that circle its head are half veiled in cloud and rain, and waterfalls are making long white snakes down the crags to where they vanish in the woodlands below. There are many arctic-alpine plants on these rocks, and in the woodland below there are many ash trees and some very large hawthorns growing on a band of lime-rich volcanic rocks. A part of this woodland, which has an undergrowth of many different calcicole plants, is now fenced in and the vegetation, free from grazing for the first time in many centuries, has responded by growing tall and luxuriant, with many seedling trees springing up. A feature of these wooded rocks, and one unique in upland Snowdonia, is a splendid display of the wood vetch (*Vicia sylvatica*) that hangs masses of its showy purple-veined creamy flowers down the ledges. The ecology of this ashwood can be compared and contrasted with the similar rich ashwood of Craig y Benglog on Rhobell Fawr. Cwm Glas Crafnant should not be confused with Cwm Glas on Snowdon.

Rhinog. The Rhinog range forms that exciting skyline of block-like mountains you see on your left as you go north from Dolgellau to Trawsfynydd. The reserve is in two parts, one north, one south of the Pass of Ardudwy. Both parts contain barren mountain, much rock and dense heath; and where the ground is less rocky it becomes wet and peaty over large areas, the sort of country that ecologists call blanket bog. This is undoubtedly the least frequented of these five mountain areas, the most intractable agriculturally, the most desolate and uninhabited, and the least rich in flora and fauna, its extreme acidity being only relieved by an occasional less acid patch of bog where lesser clubmoss and a fair variety of sedges are found. The luxuriance of the heather on these Cambrian rocks is presumably due to the acidity and low fertility of the soil, and the smaller number of sheep per acre than the average kept by farmers in Snowdonia. It seems likely that the wide, grassy slopes of Snowdon, Cader Idris and elsewhere are so free from heather partly because the sheep keep it down, partly because their soils are more fertile than those of the Rhinog. If you follow a forestry fence you can often see where heather grows thickly inside it but is dwarfed in the sheep-nibbled turf outside. A very interesting plant of the Rhinog, where sphagnum moss grows among the heather, is that tiny but attractive orchid, the lesser twayblade, which in some years is abundant there.

Woodland Reserves

The seven purely woodland reserves in the Park differ greatly one from another, which is why they were chosen as reserves. Between them they represent practically every type of deciduous woodland in the Park, from the most nearly natural oakwood (Rhygen) to the mixed type of woodland deliberately modified by landowners to embellish their estates (Camlyn). They also represent woodland growing on many different kinds of soils from very acid (Ganllwyd and Cymerau) to base-rich (Tremadoc). There are woods that are dry underfoot (Tremadoc) and those which have marshy patches (Gorswen). There are woods which have what is for North Wales a moderate rainfall (Gorswen 50 inches), and woods of heavy rainfall (Ganllwyd and Cymerau 80–90 inches). Most of these woods have for centuries been grazed by sheep, as is the custom in upland Wales. On becoming reserves they were fenced, and the effects of this exclusion of grazing animals are being closely studied.

Gorswen. This is a charming stretch of gently sloping

woodland in a beautiful lowland area of the lower Conway valley near Ro Wen. It is a type of woodland not commonly found in Snowdonia, having much more the feel of a pedunculate oakwood of Shropshire or Herefordshire. There are in fact more pedunculate than sessile oaks in it. Many of the wild flowers, too, are those of base-rich woods: sanicle, dog's mercury, herb robert, yellow archangel and slender false-brome grass (*Brachypodium sylvaticum*). But apparently the soil is not quite rich enough for such species as wood spurge, wild daffodil and Solomon's seal. In Gorswen I listed more plants and birds than in any other of the woodland reserves; and insect life, too, seemed particularly plentiful. This wealth of life is a reflection of the good soil (glacial drift from base-rich volcanic rocks higher up) and the many little streams oozing through the wood and spreading marshy patches along their courses. These base-rich woods are the typical haunt of the marsh tit in Snowdonia; so much so that if one could speak of birds as calcicoles and calcifuges I would call the marsh tit in Snowdonia a calcicole, and the far more widespread willow tit a calcifuge, for it is characteristic of wet woods on the most acid soils.

Dolgarrog. A couple of miles from Gorswen, Dolgarrog wood is unique in being an oakwood growing on those basic pumice tuffs which on the mountains are the rocks that bear the richest arctic-alpine flora. These calcareous rocks form the massive buttress in the centre of this very steep wood that climbs nearly 800 feet above the Conway flats and, final outpost of the Ordovician rocks, looks east across the smooth Silurian landscape beyond the estuary. As other rocks in this wood are acid rhyolites, there is a mixture of lime-tolerant, and lime-hating plants. Those on the calcareous rock include woodruff, sanicle, ramsons, wild strawberry and soft shield fern. The main calcifuges are foxglove, bilberry and bracken. An attractive tongue of alderwood extends towards the moorland from the top edge of this wood, following the stream called Afon Ddu. This is a rare relic of what was once a common type of woodland, known to have been widely cleared in the sixteenth century to create semi-upland pastures when the fashion arose to graze store cattle on the lower hills (the age of the beginning of the drovers' roads). Probably these alders above Dolgarrog survived because of their inaccessibility. A typical plant growing under them is the marsh hawksbeard, a species common on the richer wet soils in this part of Snowdonia but much rarer further south in Merioneth, where the best locality I know for it is in a similar extremely wet alderwood near Rhos-y-gwaliau. I presume some of these little alderwoods have survived because they supplied wood for clog-making.

Coed Tremadoc. The fine scarp-face of high cliffs rising behind the trim village of Tremadoc is popular with rock-climbers because the dolerite that forms it has so shattered as to leave tall, hard, vertical faces. It is also this dolerite (with some pumice tuffs) which, being mineral-rich, makes these rocks attractive to several species of lime-loving plants. Among the wild chaos of immense boulders under the cliffs grow hartstongue fern, marjoram, wild privet, spindle, orpine, shining cranesbill and rock stonecrop. Oak, ash and wych-elm have sprouted between the boulders and some cling desperately to the cliffs. In summer many of the high ledges are beautifully massed with ox-eye daisies. This reserve has a large jackdaw colony; there are also ravens, kestrels, sparrowhawks and many smaller species. A few choughs roost in the cliffs in winter. The block-scree is reputedly a haunt of pine-martens, as well it might be, so inaccessible are the caverns below it. There are red squirrels and several badgers' sets, one of which is said to be ancient.

The remaining four woodland reserves, all on the very acid soils, can be considered as a group. The highest is Coed y Rhygen, reaching 800 feet on the shore of Trawsfynydd Reservoir and very close to the atomic power-station. It is a wild, bouldery wood of sessile oaks on Cambrian rock. Trees and boulders alike are green with mosses and feathered with lichens. Polypody ferns bristle from the branches. It is a strange and beautiful place reminiscent of Wistman's Wood on Dartmoor, except that Wistman's consists, surprisingly on such acid soil, of pedunculate oaks. Possibly the rise in temperature of Trawsfynydd Reservoir due to its water being used by the power-station may modify the climate of this wood and bring about vegetational changes.

The preservation of Coed Ganllwyd, another high-rainfall wood eight miles south of Rhygen, would have pleased the early travellers, for in this wood is Rhaeadr Du, one of the three waterfalls they all visited in the Mawddach region. Here too came the fern-collectors, for ferns are abundant in the shade and spray of this fall. And it was a happy moss-hunting ground of D. A. Jones of Harlech, in his day the leading Merioneth botanist and one-time president of the British Bryological Society.

The two remaining woods were also known to the tourists of long ago, for Coed Cymerau and Coed Camlyn are part of the famous woodlands that clothed the Vale of Ffestiniog. Cymerau's sessile oaks cover both sides of a stream that hastens down a deep valley. The air is often moist with the spray of waterfalls that cascade into a green world of ferns and mosses. It is the home of fox, badger and otter; of raven, buzzard, pied flycatcher and wood warbler. A similar fauna inhabits Coed Camlyn farther down the valley. This wood stretches for about a mile on a steep slope facing north-west, opposite the old mansion of Tan-y-bwlch whose former owners planted many of these Ffestiniog woodlands. For many years before this wood became a reserve not much growth of seedling trees had been possible because of sheep-grazing. From now on natural regeneration should be assured. But rhododendrons, so often the curse of woodlands on acid soils, are abundant in Coed Camlyn, probably being introduced years ago as pheasant cover. Rhododendron is difficult to get rid of because it so readily sprouts from the stumps. Coed Camlyn is one of the few Snowdonian woods where all three British woodpeckers breed, for the lesser spotted, though lately increasing, is still rather rare.

Several other fine stretches of deciduous woodland remain in the Vale of Ffestiniog and there are still others in neighbouring valleys. Nearly all such woodland can be stigmatised as 'uneconomic' and is therefore under the threat of being replaced by conifers, especially these days when private enterprise is finding afforestation a good investment. If there is going to be much more clearing of deciduous woodland in North Wales, there will certainly be a severe reduction in some of the Park's most characteristic birds. The buzzard, for instance, so much a part of the deciduous hillside woods, must be gravely affected.

Shore and Dune Reserves

Finally, there are the magnificent dunes of the north Merioneth coast that are so rich in wild plants, insects and birds that every naturalist would wish that these two reserves of Morfa Harlech and Morfa Dyffryn extended much further than they do. 'Morfa' means literally 'a sea-place' and is a common word for flat land along the coast. Often such low-lying land has built up on storm beaches. Morfa Harlech, for instance, developed when shingle banks grew northwards at a sharp angle from the now inland cliffs of the old coastline. Morfa Dyffryn is almost the same triangular shape and has built up in the same way. Dry sand blowing inland off the beach at low tide has piled up into large dunes, some loose, some fixed by vegetation. Just as on the mountains, so in the dunes there are acid areas and lime-rich areas: sand without shells being acid, sand mixed with crushed shells being calcareous. The other vital factor for plant life is water: and although dunes look very dry, they usually are dry only on the surface and contain plenty of water inside them. Hollows in dunes are often almost permanently wet even at the surface and can be shallow pools for long periods during a wet winter. The calcareous wet hollows are rich in plants such as creeping willow, marsh helleborine, various sedges, early marsh orchid and dwarf purple orchid. The sharp rush (*Juncus acutus*), which is very fine in these dunes, was first recorded for Britain here by Thomas Johnson in 1639 on his way back to

PLATE IX **Mountain-top Erosion.** *Above*, the shattered rocks of the Castle of the Winds on the top of Glyder Fawr, with Snowdon in the distance. *Below*, Cwm Idwal, showing boulder-scree fallen from the cliffs around the Devil's Kitchen. Note also the downfolded rocks (syncline) of these cliffs (see page 45) and, down by the lake, the hummocks of glacial moraines

PLATE X **Butterflies.** *Above*, the marsh fritillary, here at a bog-bean flower, is local and usually found on peat bogs. *Below*, the small pearl-bordered fritillary is more widespread and is found in moorland valleys, woodlands and marshy places

PLATE XI **Moths.** A newly emerged male emperor (*above*) with its cocoon. It is a typical heather moor species and is on the wing in early spring. The scarlet tiger (*below*), rare in most of Snowdonia, is found mainly in south Merioneth

PLATE XII Snowdonia's rarest mammal, the pine marten, an agile tree-climbing member of the weasel family, lives both in forestry plantations and on open rocky mountains. Mainly nocturnal, the marten is very rarely observed

PLATE XIII The polecat (*Top*) is plentiful in Snowdonia especially in lowland and semi-upland districts. It interbreeds readily with escaped ferrets, the resultant hybrids usually showing more white in their pelage than is found in wild polecats. The red squirrel (*above*) is a much decreased species and in many areas of the Park has been largely replaced by the grey squirrel

PLATE XIV *Above*, Moel Siabod from the north-east, rising from the valley of the Llugwy. *Below*, Craig yr Ysfa on the Carneddau, photographed from Pen yr Helgi-du. The volcanic cliffs of Craig yr Ysfa are well known to both climbers and botanists

PLATE XV Aberglaslyn Pass. The Scots pines are a nineteenth-century addition: an eighteenth-century drawing shows the gorge devoid of trees

PLATE XVI *Above*, Cnicht from the south west and the River Croesor near Llanfrothen. A long, narrow ridge, Cnicht appears as a sharp peak from this angle only. *Below*, a farming scene near Towyn, Merioneth. Working horses are now uncommon but have not been entirely ousted by the tractor for tillage on the steepest slopes

England from Snowdon. The drier slopes of the fixed dunes are fragrant with many common but beautiful species such as thyme, lady's bedstraw, burnet rose and rest-harrow, and there are typical calcicoles such as hounds-tongue, blue fleabane, lady's fingers, field gentian, autumn lady's tresses, pyramidal orchid and bee orchid. Two very rare plants of this coast are the pendulous helleborine and the minute spike-rush (*Eleocharis parvula*). There are many nesting birds: among them, stock doves, ringed plovers, redshanks, oystercatchers, peewits, snipe, black-headed gulls, meadow pipits, reed buntings and skylarks. And in winter on the pools and marshes there is often a variety of ducks as well as whooper swans. In summer the insects are many and varied. There are dragon-flies around the pools; sometimes vast numbers of common blue butterflies; dark-green fritillaries; cinnabar and burnet moths; and in occasional years large numbers of poplar hawk-moths which have fed as caterpillars on the creeping willow. Though the sea campion flourishes well here, most of its seeds are eaten by the larvae of several moths of the genus *Hadena* such as the campion and lychnis moths. In some years there is a caterpillar or two in nearly every capsule.

The open shore, all the way from the Dwyryd to the Ysgethin, is a fine sweep of sandy coastline, varied, as at Mochras (Shell Island), with boulders and shingle. The plants of shorelines are there—saltwort, sea rocket, sea kale (very rare here), sea sandwort and several oraches. There is also the fascinating natural litter of shorelines: whelk egg-cases, an occasional dead pipe-fish, sometimes a dead oiled sea-bird, and at rare intervals wrecks of small wind-drifted jellyfish such as by-the-wind sailors and Portuguese men-o'-war.

For a final word about Snowdonia's nature reserves, let me go back to Morfa Harlech. Morfa Harlech nature reserve with its flowers and birds and butterflies is a place of much delight, a place where you can experience vividly all the colour, the scent, the vitality of wild nature. Yet it is certainly a mere fragment of a former wilderness of saltings and creeks, pools and fens, that covered several square miles of this morfa centuries ago before the sea-banks were built and the drainage engineers and lately the Forestry Commission laid their hands on it. So it is with many of these nature reserves: they are remnants almost miraculously salvaged from the Snowdonia of centuries ago when oak forests were vast and full of deer, and large birds of prey were everywhere abundant; when water and fenny wastes were widespread and marshy river-courses filled the valley bottoms. Something of that world survived as late as the eighteenth century but now for 200 years natural habitats have shrunk so rapidly that the remnants that are left are of priceless value. And I do not mean valuable to biologists only. Some biologists may see nature reserves as mere outdoor laboratories (which one suspects they would gladly take indoors if they could). But nature reserves are much more than that. They are, by definition, reserved for nature not for man. They are places which man with a rare spirit of unselfishness has set aside so that the wild species of the world can survive instead of being utterly exterminated. We must never think of them primarily as reserves for scientific research: they are that only incidentally. But there is a place for man in nature reserves, not only for scientists and naturalists but for those who will visit and not disturb, and for whom there is enormous refreshment of the spirit to be got from being for a while in a place of genuine wilderness and natural beauty, to enjoy what Thoreau defined as 'the tonic of wildness'.

The Nature Conservancy

In these already existing nature reserves, along with some other areas which would also make excellent reserves, you can see a representative cross-section of the great wealth of Snowdonia's natural habitats from sea coast to mountain top, and the wild plants and animals that are characteristic of them. You may see and admire and go away happy in the thought that something is being done to safeguard wild life in a world that gets increasingly hostile to it as man, armed with ever more powerful tools, attacks the countryside with increasing destructive powers, but not, alas, with equally increasing knowledge of what he is really doing.

It is to the search for such information that much of the Nature Conservancy's attention is directed, and necessarily so, for successful nature-reserve management calls for a deep knowledge of the relationship between plants, animals and their environment. Establishing a nature reserve is far from being a mere matter of fencing sheep, and possibly people, out and then letting nature rip. For you may soon find it invaded by undesirable plants and animals that overwhelm those species you particularly want to protect. Or you may find that in your reserve one very desirable species is elbowing out some other very desirable species. What can be done in such circumstances? Often the answer is not readily available. Nature conservation is a new activity for mankind and we are still very ignorant about what controls the often delicate balance between this and that species. What we can safely predict is that the causes of any natural phenomenon will turn out to be multiple and complicated. And since all life springs from the soil, it is clear that it is the study of the soils of nature reserves, and their relationship with the animals and plants, that forms the most vital approach to all problems of reserve management.

For this reason soil-studies are a key part of the Nature Conservancy's scientific programme carried out at its Bangor research station, whose windows appropriately enough have a superb view of the Caernarvonshire mountains. It is the grasslands visible on those vast slopes that have received much of the Conservancy's attention in the belief that in these grasslands is hidden the answer to many fundamental problems. The grasslands, though they look very uniform from a distance, are in fact very diverse, not only in species, but in nutritional value, a diversity which is partly due to the many different igneous and sedimentary rocks on which they grow and partly to differences in rainfall or sunshine or temperature or aspect. One way of studying the various nutritional values of grasses is to study the sheep that graze them. Most hill-walkers would probably think that sheep graze almost everywhere equally; but in fact the sheep have very decided preferences. They know where the bite is sweetest and it is interesting that their choice in feeding-places can often be correlated with the nature of the rock below the grass. What is remarkable is that the sheep are sensitive not merely to the difference between, say, mat-grass and agrostis-fescue grassland; they may even be sensitive to the amount of trace elements, such as cobalt, in the soil. But there is no absolute proof of this.

But rocks and soils have important physical as well as chemical differences between them. These, too, are being very closely studied, an instrument especially useful to this end being the X-ray fluorescence spectrograph, a machine able to analyse quickly and accurately the mineral elements in rocks and plants. By detailed study of this kind the ways in which soils really influence the nutrition of plants are being investigated. In pursuit of these studies the Nature Conservancy has not confined its attention to its nature reserves but is looking at Snowdonia as a whole, and the soil and vegetation of the mountains has been mapped in detail. One striking aspect you see at a glance from these maps is the rich diversity of soils west of the River Conway on the Ordovician formations, with their mixture of volcanic and sedimentary rocks, compared with the comparative uniformity of soils east of the Conway on the entirely sedimentary Silurian rocks.

Not that grasslands and nutrition are the only objects of the Conservancy's research. There are also studies in hand of the arctic-alpine plants of Snowdonia: why they are there, what decides their habitat preferences, what is the effect on them of competition, grazing, climate and other influences; and what could be done to rehabilitate struggling and rare species by, for instance, limiting soil erosion. Many of the fences erected near the top of Crib Goch are connected with such experiments.

In connection with its work of research and reserve management the Conservancy maintains a team of wardens who not only protect the reserves but also give lectures about them, conduct parties over them, and in various other ways help to instruct the public about the reserves and the need for conservation.

To sum up: what all this long-term fundamental research is aiming at is to find out exactly what happens to nature when you set up a nature reserve. In a way the cart has been put before the horse. Ideally one ought to find out all the answers and then create nature reserves when you know exactly how to manage them. But wild habitats are disappearing so quickly that we have now long since passed the point when there is any course left but to try to save what few natural or semi-natural places still exist, and hope that we shall learn in time the best methods of looking after them. As E. M. Nicholson has said: 'Our starting point then must be to recognise with humility that whatever we do must be based more on ignorance than on knowledge and that some at least of our interventions are likely to prove wrong. Nevertheless experience with nature reserves shows that nothing can be more wrong than doing nothing when a habitat is rapidly deteriorating.'

CHAPTER SEVEN

Carneddau, Glyder and Snowdon

I begin this descriptive half of the book near the northern tip of the Park at Caerhun. And what more delectable spot for a beginning than these tree-studded fields not many feet above sea-level where the Romans placed an important stronghold to guard the tidal waters of the Conway river? When choosing their chief Welsh sites the Romans had an instinct for combining strategy with comfort, and here they had sunshine, a mild climate and shelter from most winds. They had unending supplies of fuel, timber and stone. They could get trout, salmon and shellfish from the estuary; and, like the British before them, they got pearls from the Conway mussels. Probably at that time the migrating salmon and sewin were encircled by nets paid out from small boats just as it is done there today, and just as it had been done for ages before the Romans. There is much else of life in the Conway valley that has changed very little. It is true there is a modern aluminium works and a hydro-electric power-station. But it is also true that after sixteen or seventeen centuries the banks of the Roman fort still boldly mark their neat rectangle on the fields. The Roman buildings have gone, a medieval church has come into one corner of the rectangle, but what else has altered? Still along the valley, despite the reclamations of centuries, you can find sizeable remnants of the former wilderness that belongs to often-flooded, alluvial ground—reed-beds, rushy levels, fresh and brackish creeks and willow thickets—places much more attractive to the naturalist than the farmer. Collectors of water-beetles and dragon-flies seem to have been particularly happy here, judging by the frequency with which 'the Conway flats' appear in the records. The estuary, with its ever-changing tides, is a place of unfailing interest for the bird-watcher, especially during the spring and autumn passage, and in winter when the wigeon and other duck are here. For the botanist there are all the varied plants that belong to estuaries and their margins. There are salt-marsh and fresh-marsh species almost side by side. There is a wealth of rushes, sedges and pondweeds; and here, too, is the only locality in Snowdonia for the four-feet tall, handsomely arching galingale (*Cyperus longus*), which was not discovered until 1946 and may be a garden escape. The large and showy pink balsam called policeman's helmet, introduced to Britain from the Himalayas and still elbowing its way along the banks of Snowdonian rivers, has forced its way from the Conway and stationed itself conspicuously on both sides of the road between Trefriw and Dolgarrog.

The Conway valley up to Betws-y-coed is perhaps as good a butterfly district as any in the Park, though it cannot be as good now as it was before conifers ousted so many hardwoods and were planted in so many open spaces. Apart from the very commonest species, I have seen the following there: holly blue, green hairstreak, large skipper, gatekeeper, speckled wood and comma; pearl-bordered, small pearl-bordered, silver-washed and dark-green fritillaries; but no marsh fritillary though it is locally common elsewhere in the Park. Nor have I seen a brimstone there, but this is not surprising, for both the buckthorns, its food-plants, appear to be absent from the valley. Besides these butterflies that I have seen myself, there are records from the lower Conway valley of brown hairstreak, purple hairstreak and dingy skipper. The rare Duke of Burgundy fritillary is on record for Deganwy at the mouth of the Conway just outside the Park boundary. The wych-elms might profitably be searched for caterpillars of the white-letter hairstreak and the large tortoiseshell, for both have been recorded in the district in recent years. But both are rare.

For certain woodland birds the Betws-y-coed area is an important centre of distribution. The woodcock, for instance, probably nests thereabouts more numerously than anywhere else in the Park, for in Wales it has a decidedly eastern distribution, breeding also around Bala. On the western side of Snowdonia it is probably no more than an irregular nester. A few woodcock lie up in high forestry plantations until well into the spring, giving hopes that they are going to nest, but seldom do so up there. Not that breeding woodcock are nearly as likely to be detected as that other crepuscular bird, the nightjar. Every observant person recognises the reel of the nightjar, which is far more widespread in Snowdonia than the woodcock, but few people seem to know the squeak and grunt of breeding woodcock. The conifer plantations round Betws-y-coed have become a home for a number of new birds, the most noticeable being the blackcock, whose strange bubbling song has become a familiar sound among the newly planted trees during the last twenty years. The lesser redpoll too has increased enormously since the coming of the Forestry Commission. Crossbills have bred sporadically, especially in the seasons after invasion years. I saw a family party by Llyn Bodgynydd in June 1963, a year in which crossbills bred widely in Britain. But the most striking newcomer is the siskin, which in Britain has always been mainly a Scottish and Irish breeding species but which has probably been nesting in plantations north and south of Betws-y-coed since the mid-1940's and is now well established. It seems to be uncommon as a breeding species elsewhere in Snowdonia but as it is not a conspicuous bird it may have been overlooked. It probably now breeds in Coed-y-brenin and in Dovey Forest; and though nesting in tall trees it often comes to the ground for it is very partial to dandelion seeds. This attractive little finch has been breeding for some years around Lake Vyrnwy in Montgomeryshire and has lately got into Staffordshire. So it looks as if we can prophesy that the siskin is on the way to getting established in many other conifer forests of England and Wales.

I mentioned the churchyard that occupies a corner of the Roman fort at Caerhun. It is worth visiting for its yews. One of them, a single trunk, took three times my outstretched arms to get round it, which gives it a girth of sixteen and a half feet. What age you give such a tree depends on how fast you reckon yews grow on good ground. And this is good ground. You have only to look at the thick-stemmed oaks just outside the perimeter of the fort. One of them, abounding with hybrid vigour, girths at an arm's length longer than the biggest yew in the churchyard and still looks youthful. Even the elders here are outsize: one I measured at twelve inches from the ground was six feet round. And, loveliest of trees, the wych-elms here are magnificent.

Not that good trees in the lower Conway valley should surprise anyone, for it was the area round Betws-y-coed that was formerly occupied by the Gwydir forest, a mainly oak forest varied with beech, sweet chestnut, spruce, birch and rowan. Most of this old deciduous forest has now disappeared and been replaced by conifers, and although spruces are beautiful climbing up rocky mountainsides with their sharp points black against the sky, most naturalists will turn with relief to the steep slope that takes you out of Betws-y-coed up towards Llyn Elsi, for that way you go through a remnant of the old Gwydir forest, a wild, cool, north-facing rocky wood of oak, ash, sycamore, hazel, small-leaved lime and even a scattering of yews. Yews can hardly be reckoned native in Snowdonia as those on the Denbighshire limestone are so obviously native, but they thrive well in the richer soils round Betws-y-coed and have presumably been planted by birds carrying up the seeds from local churchyards or estates. In these woods, in contrast with the neighbouring conifer plantations, is a wealth of forest-floor plants—wood sanicle, enchanter's nightshade, wall lettuce, twayblade and many other species, and, as is natural in a shady, damp wood, an abundance of ferns. But as you go higher this vegetation gets sparser and heathier, as happens in most steep woods.

The Romans had a nose for mineral wealth—it was presumably their main object in occupying Snowdonia—and it is no coincidence that within easy reach of Caerhun there are numerous lead-mines. I suppose most people recoil from these old mines with distaste when they see the derelict buildings, the rusting machinery, the dangerous shafts, the gaping levels and the poisonous vomit of grey spoil left after the processing of the ore. Yet as I have said in chapter 2, they are fascinating places for a naturalist though they face him with formidable ecological problems. Many of these mines burrow into the hanging slopes, now afforested, that drop several hundred feet along the west side of the Conway between Betws-y-coed and Dolgarrog. If you make your way up these advance ramparts of the Carneddau you find yourself on a shelf of country at about 800 feet above the sea, a country utterly different, a strange, sparsely populated world of far-scattered houses and lead-mines that year by year are getting more and more isolated and hidden among the growing conifers. Here, too, is a small lake-district of natural pools and mine reservoirs, some of them deep and rock-girt and looking quite alpine amid the spruces and pines; others are broad, shallow, half-empty, peaty lagoons with wide, muddy margins. It is a district where you may look for the plants often associated with lead-mines such as alpine pennycress and forked spleenwort. And here, covering a few square feet in one of the bogs, is the prostrate, apple-green marsh clubmoss growing in a mat of white beak-sedge. This clubmoss, which in Britain is least uncommon in the south of England, is extremely rare in Wales, this being almost its only known present locality. Round the pools there are a few wading birds such as peewits, snipe and sandpipers breeding; and there are little grebes and black-headed gulls. This is a region in which it would be rewarding to study several branches of natural history because of its contrasting habitats. For there is not only the quick change in height from the Conway river up to this shelf and the change from deciduous to coniferous woodland; there is also the variety afforded by meadows, heaths, lakes and marshes and by the differing degree of acidity in the ground; and there is the contrast between relatively undisturbed soils and those belched out of the mines.

By way of Roman lead-mines we can grope our way back into the shadowy world of prehistory: for doubtless the Romans merely took over long-existing Celtic mines, workings contemporary with the construction of the Iron Age hill-forts so generously scattered about the perimeter of the mountain region. But human settlements above the Conway are far more ancient than that. We can assume a continuity of life backwards from the hill-forts through the Bronze Age to the megalith builders, who are reckoned to have migrated to the Irish Sea region from the Mediterranean during the centuries around 3000 B.C. Their stones, dolmens and circles still stand along this wide shelf and along the corresponding shelf east of the Conway where, in the small Denbighshire portion of the Park, you can see the chambered tomb of Capel Garmon. There, perhaps more easily than anywhere, when you see the wide and wonderful view of the Caernarvonshire mountains from that sweet-smelling, healthy platform on the flank of the Hiraethog moors, you can begin to feel your way towards some understanding of Neolithic life.

For Bronze Age man we can go five miles west of the Conway to the vicinity of Llyn Dulyn where, on a south-facing slope at about 1,700 feet, are the copious Dartmoor-like traces of hut-circles and cairns and a bewilderment of walled enclosures and what look like terraced fields. This valley-head right under the Carneddau may strike you as a sodden, lonely and unkind place to have settled to till the earth. But if in Bronze Age days the climate was warmer and drier than to-day's, then the upper Dulyn valley could have been a delightful habitat for man even in winter. The name of this spot is Pant-y-griafolen—Rowan-tree Hollow, a name, perhaps a very old name, that recalls trees that have long since gone: for apart from a clump of planted conifers there is not a tree in the place, only wide and empty grassland climbing away to the horizon. But is the grassland so empty? These high grasslands, we should remember, have never been ploughed and, especially near ancient settlements, they may be repositories for much precious evidence that one day may come to light about the life of Bronze Age man and the natural history of those days.

For a glimpse of the life of Iron Age man go up from Llanbedr-y-cennin to the breezy, commanding height called

Pen-y-gaer, a hill-fort with stone circles, substantial walls and banks, and an elaborate defence of pointed upright stones unique in England or Wales. This 1,200-foot hill is a magnificent place from which to appraise what man has made or has not made of the countryside thereabouts in recent centuries. Below you to the north is the gentle-looking slice of rural Welsh civilisation in the fertile saucer centred on Ro Wen. It is made up of a few delightful lowland square miles west of the Conway estuary, a place of many fields coloured in summer with varied shades of green and brown, divided by countless hedges, the slopes of the saucer almost closed round with patches of sloping woodland, one of which is now the Gorswen nature reserve. Then, to the west of Ro Wen, the contrast: the pale green hills at first tree-dotted, then open grassland climbing up and away, divided into enormous enclosures by long snaking walls that go far up the slopes. Then, beyond and much higher, the lofty Carneddau, smoothly undulating in the north but becoming increasingly rugged and crag-sided in the south at Craig Dulyn and the great cliffs of Cwm Eigiau. It is all a fair country, but badly marred by the lines of electricity pylons going over the ancient pass of Bwlch y Ddeufaen.

The southward prospect from Pen-y-gaer slopes first across fields that went out of cultivation centuries ago but whose outlines are still traceable on the land. Beyond them is the deep gash of the Dulyn ravine, a delightful cool gorge with a bouldery stream and waterfalls, overhung by oaks that seem to grow out of the living rock; a home for shade-loving plants and ferns, and buzzards, dippers, grey wagtails and pied flycatchers. Beyond the Dulyn stream we are back to the shelf above the Conway. But this part is almost completely depopulated. There are a few ruined farms; and the forlorn words on the map: 'Ardda Village (remains of)' in archaic lettering.

Symbolic of the life of the past seven or eight centuries on what was a rather remote upland are the old churches that still survive with a few or no houses near them. Go to Llanrhychwyn, for instance, to see a small and simple church that grew entirely, I presume, out of local materials:

its walls look like the local dolerite; its oak timbers were no doubt Gwydir oaks; and the skill that shaped and joined them was doubtless local skill. It is a church alone in its fields, the tide of human life having ebbed. Now it stands in great peace under its arching yews. And just as the yews of Betws-y-coed churchyard have probably spread their descendants into the woods above, so the yews of Llanrhychwyn church are the most likely parents of yews that now grow on the doleritic rocks in the neighbourhood. Behind the church are old meadows, the sort you can still find in fair quantity in Snowdonia in the 800–1,000-foot region, very flowery meadows and pastures on the better soils which have never been ploughed or 'improved'. They still yield an annual crop of richly mixed herbage, which may be uneconomic but is probably a healthier diet for animals than modern leys produce.

I mentioned as conspicuous in the view from the hill-fort of Pen-y-gaer the black crag that hangs above Llyn Dulyn, a crag which, like most others in Snowdonia, is the result of the shattering effects of glaciers and frost It is on rocks like this Craig Dulyn or its neighbours Craig yr Ysfa and Creigiau Gleision that botanists find the best samples of the mountain flora of the Carneddau's eastern valleys. Not that many botanists seem to go there. Most who come to Snowdonia are lured by more famous cliffs further west on Snowdon itself or on Glyder. Or they make their way to the cliffs of Ysgolion Duon on the western side of the Carneddau, floristically very rich cliffs which Pennant called 'the most horrid precipice that thought can conceive'. But Pennant was not a botanist.

Craig Dulyn, a fine cliff rasped out of the flank of Foel Grach, is a perfect example of a crag overhanging and partly encircling a deep mountain lake. Llyn Dulyn, the Black Lake, is genuinely black no matter how blue the sky. Seeing its evident great depth, its almost perfect circular form and the cup-shape of the rocky hollow in which it lies, you can very easily understand how people mistook such lakes for craters when the idea of the volcanic origin of Snowdonian rocks was first being mooted. Its plants are the normal mixture of lime-lovers and lime-haters and those which are indifferent; and as usual there are alpine and lowland species all enjoying mountain life equally well. It is, as elsewhere, along the watercourses that you find the plants most abundant and a special delight of Craig Dulyn is the luxurious quantity of the chickweed willow-herb (*Epilobium alsinifolium*), a very attractive species with its hanging, deeply purple-pink flowers. Yet though it thrives so well here, this is almost its southern limit and in all Britain there is not a suggestion of it farther south than Snowdon. But plant and animal distribution is full of such sudden, inexplicable full stops.

From the Dulyn rocks it is a short climb to the backbone of the Carneddau, the magnificent ridge that undulates over far-spreading grasslands for many miles from Y Drum in the north over Foel Fras, Foel Grach, Carnedd Llwelyn, Craig Llugwy, then above the north-facing precipices of Ysgolion Duon to Carnedd Ddafydd and Pen yr Oleu Wen. In spring it is a place of meadow pipits, wheatears and singing larks, an unbroken chain of lark-song stretching from end to end of the Carneddau. But on the open dry grassland there are very few other species. This grassland will not excite the plant-seeker either unless he is a specialist in such habitats, but there are some very interesting problems concerning the ecology of these high pastures, as I have mentioned in chapter 6. All the summit rocks have a similar flora: the woolly-haired moss in great abundance; the lichen miscalled reindeer moss; the stiff sedge, the alpine and fir clubmosses; bilberry, cow-berry, crowberry and sometimes the least willow. Though on a sunny day it is very lovely to see the cloud-shadows playing endlessly across the great spaces of the Carneddau and though, even today, you find peace and solitude there, yet it is difficult not to feel beckoned by the exciting sweep of wild summits and crags you see beyond Nant Ffrancon stretching all round the south and west from Gallt y Ogof through Glyder and Tryfan to Carnedd y Filiast.

The richly varied habitats included in the Carneddau region are spread over many square miles of mountain and valley. In contrast the Glyder range, being wholly an upland area, has far less variety but is much more compact. The best botanical localities on the Glyder are far more easily accessible than those on the Carneddau. Llyn Idwal is less than a mile from a main road. But to get to Ysgolion Duon means a longish trek up the Llafar valley from Bethesda. The Glyder peaks are lined up conveniently along the Bethesda-Capel Curig road in a beautiful seven-mile curve, with an almost symmetrical pattern of east- or north-facing craggy hollows gouged out of the slopes the whole distance. All have their interest, but Cwm Idwal and Cwm Bochlwyd with their lime-rich rocks have for centuries been recognised as the great localities. So it is round the edge of one or other of these lakes that most naturalists find their way up to the summit ridge. Not until you get up there do you realise what a narrow, roof-like range this is, for all along it on the west the land slopes immediately down to the Pass of Llanberis and there is little in the way of a summit plateau except in the few saddles between the peaks.

Animal life, especially vertebrate life, is naturally sparse along so exposed a ridge. I have seen common shrews, field voles and a fox and there are probably pygmy shrews and hares. The dead hedgehog found here at 2,500 feet, (recorded by Forrest) was surely carried up by buzzard or raven. Frogs and newts are quite usual at such heights, breeding in lakes and smaller pools and making their way into all sorts of mountain habitats after the breeding season. I have seen a newt on a wet mossy ledge halfway up the Devil's Kitchen. Charles Oldham saw a snow bunting on Y Garn at 3,100 feet on 4 April, 1913. At such a date perhaps it gave him faint hopes of becoming the first to record this species breeding in Wales. But that happy day has still not arrived.

On a summery day it is the insects that are more conspicuous than any other creatures at over 2,500 feet. Coming down the long grassy slope from Y Garn to Llyn y Cwn on a late July day I noted two butterflies, the resident small heath and the migratory large white; there were many flying ants; a few froghoppers, spiders and grasshoppers; a copious hatch of small craneflies in several damp areas of turf; a courtship flight of gnats 'singing' with a very high-pitched whine at the lake-edge (the 'fairy music' of the lake legends?); ants, woodlice, spiders and beetles in the scree above the lake; in the rocks a light-grey species of carpet moth was common; the common yellow muck-fly rose abundantly from the sheep droppings; and in the late evening when I reached the top of Glyder Fawr, the air was very still and warm and I was quite as tormented by midges as I would have been 2,000 feet lower.

Lying at 2,500 feet at the foot of the screes of Glyder Fawr, Llyn y Cwn is interesting as one of the highest lakes in Wales. It has one edge rocky, the other 'marish', to use a word Leland was fond of when he found his boots sinking into the margins of Welsh lakes. It is a type of vegetation common round moorland tarns. The mat-grass moor slopes into a squelchy area of bog moss, cotton-grass and star sedge and this merges lakewards into a zone of bottle sedge growing in a foot or two of water. The only other obvious plants are the triple leaves of the bogbean standing out of the water on its stiffly held stems; and the long, pale-green leaves of the flote-grass that lie on the surface in strange and beautiful patterns. In the stony shallows on the rocky side of the lake you can see the yellow-green flat-leaved stars of the water lobelia, some of them sending up from the centre of the rosette a shoot just long enough to hold the nodding pale-lilac flowers a few inches above the surface. In other words, Llyn y Cwn has the typical impoverished flora of acid, upland lakes. But its water is not peaty as so many such lakes are, but is very clear: which only makes it the more regrettable that people use it as a receptacle for discarded cans.

The name Llyn y Cwn, meaning the Lake of the Dogs, is only one of several animal place-names on the Glyder range. There are for instance, *caseg*—a mare; *filiast*—a greyhound; *march*—a stallion; and *cywion*—chickens. It is an interesting group of names, probably all of them memories of an ancient hunting and pastoral way of life. The dogs may have been hounds; and the chickens were perhaps young grouse, for a Welsh word for the adult grouse is *iar y mynydd*—hen of the mountain.

The Reverend John Evans developed a fanciful theory about Llyn y Cwn in order to explain the moraines in Cwm Idwal. He believed that the present Llyn y Cwn is only a relic of a much vaster sheet of water once poised above Cwm Idwal. There came the day when this great high lake burst through its impounding cliffs to create the chasm we call the Devil's Kitchen and the resultant flood strewed all the shattered rocks around Cwm Idwal. Not that we should be too hard on this idea. Evans, after all, was writing in 1798, 33 years before Darwin came to Idwal and failed to perceive the evidence of glaciation there, though when it was pointed out to him years afterwards he admitted it was obvious enough. Later on, in 1842, Darwin became the first to give full details of the glaciation of Cwm Idwal. All the same, it is only fair to point out that Evans was on the right track in sensing that there was some unusual significance in these 'streams of huge stones pointing to the hollows that lead to the chasm', and in perceiving they had not merely dropped off the precipices.

There is a well-trodden path up Glyder Fawr passing near Llyn y Cwn (hence the discarded cans). It goes up one of those tedious dry screes all too common as the chosen paths of mountaineers. How much more interesting people's hill-walking would be if they would take even a mild interest in mountain plants and make their way, not up the dry scree where little or nothing grows, but up the wet scree where there is always something worth seeing. But then, too much trampling up the wet scree would soon destroy the vegetation, so perhaps it is as well we are not all botanists. Where there is constant moisture and good drainage there is usually a fair selection of plants even though the rock may be poor in available minerals. But up the scree of reddish and whitish coarse-grained stones that slopes up from Llyn y Cwn you learn that the rock is not poor when you see butterwort and lesser clubmoss abundant near the streamside; and the shining dark-green leaves and tiny purplish flowers of the alpine meadow-rue; as well as the clustered white stars of that lime-lover, the vernal sandwort.

Halfway up, the stream and its voice are suddenly lost under the stones and the vegetation peters out abruptly. Time now to straighten your back and admire the view. Across the valley on the west the slopes are building up towards Snowdon, and back in the north to the left of Y Garn you get that great view down to Llanberis Lake, a fine, remarkably straight, glaciated valley complete with ice-worn rocks and hanging valleys along its sides. Beyond the Vale of Llanberis you see the Menai Strait, then across the flatness of Anglesey to the very ancient rocks of Holyhead Mountain with the sun gleaming on the sea beyond.

Then up the rest of the scree to reach a gentler slope on the highest shoulders of Glyder Fawr. Across a wide stretch of stones and woolly-haired moss you come to the rocky wilderness of the summit. From there, if you are lucky in the weather, you have one of the great views of this National Park. You see how the hard volcanic rocks of Snowdonia have survived the weathering of the ages while the softer sedimentary strata that once lay deeply over them have been eroded away and carried off as debris and soluble salts down the rivers. You see Siabod, Arennig, Aran, Dduallt, Rhobell Fawr, Cader Idris, Moelwyn, Rhinog. They look at each other across miles of grassy moors and heath and scree. In all that mountain landscape the valleys are hidden. You can believe it an almost uninhabited world. Then as likely as not Snowdon will breed thick clouds, for Snowdon is a very productive cloud-factory, and send them across the pass: and you are shut in with the naked rocks of Glyder and your own thoughts about time and eternity and the slow forces that will weather away even these volcanic peaks at last.

The strange primeval world of these Glyder summits is quite different from anywhere else I know. It is an expanse of loose and shattered rocks, some immense, some merely scree, set down in a vast bed of woolly-haired moss. Generally the rocks have weathered whitish, the big ones standing up as vertical columns from a vast detritus of smaller shattered stones. The pale faces of the rocks are patched by bright yellow-green lichens which are characteristic indicators of very acid rhyolitic rocks such as these. At a distance the stony plateau looks as if it is going to be painfully loose to walk on but in fact it is quite comfortably solid. You would hardly think it a likely habitat for sheep, yet even here they graze and even here they can afford to be selective in what they eat. Neatly they nip off the very sparse fescue grasses from between the stones; and they take the heads off the few plants of stiff sedge so that it is hard to find one in fruit. What they leave for you is fir clubmoss, parsley fern and the woolly-haired moss.

From Glyder Fawr you descend only a little to cross the saddle to the rocks of Glyder Fach less than a mile away. On your left you get a sudden dramatic view down the precipices of the Nameless Cwm and over more hidden drops lower down into Cwm Idwal. Scramble that way down to Ogwen if you will. If you do so you will cover some classic botanical ground and see a rewarding cross-section of Snowdonia's mountain plants. If you stay on the saddle and are content with a mere glance down at the highest ledges you will, if it is full summer, see bright pink flowers of thrift alongside white-flowered tussocks of scurvy-grass and enjoy a momentary illusion that you have strayed to the edge of a cliff above the sea in spring.

If you think after Glyder Fawr that you have seen the ultimate in the weathering of summit rocks you are in for a surprise on Glyder Fach. The pile-up of mountain wreckage you reach first and which they call Castell y Gwynt—the Castle of the Winds—is astonishing enough. But when you clamber on to it, noting on top some good specimens of stiff sedge unreachable by even the sheep, what do you see beyond but more and stranger heaps of great columnar stones ahead? One group stands on end, splaying out like a bunch of swords carelessly thrust into the ground. Another pile, the one made famous in drawings and photographs, is composed of such carefully

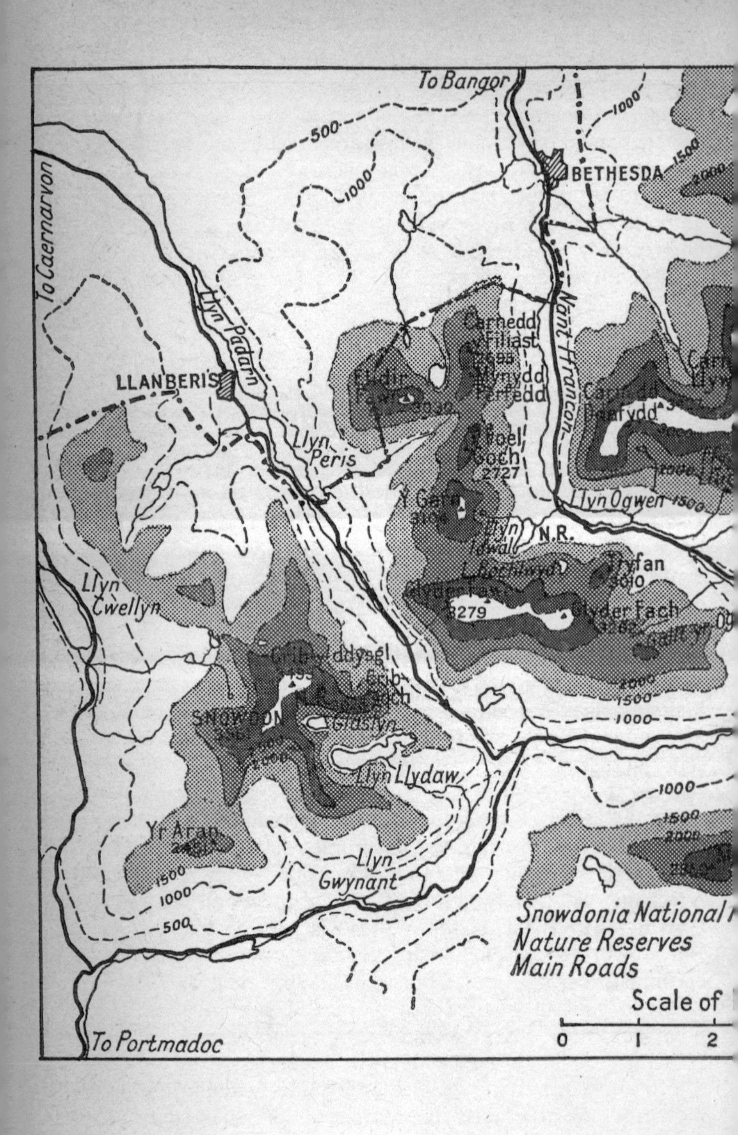

To Bangor
BETHESDA
To Caernarvon
Llyn Padarn
LLANBERIS
Carnedd y Filiast
2695
Mynydd Perfedd
Nant Ffrancon
Foel Goch
2727
Llyn Peris
Llyn Ogwen
Llyn Idwal
N.R.
Tryfan
3010
3279
Llyn Cwellyn
Crib Goch
SNOWDON
Llyn Llydaw
Yr Aran
Llyn Gwynant
To Portmadoc
500
1000
1500
2000
Snowdonia National
Nature Reserves
Main Roads
Scale of
0
1
2

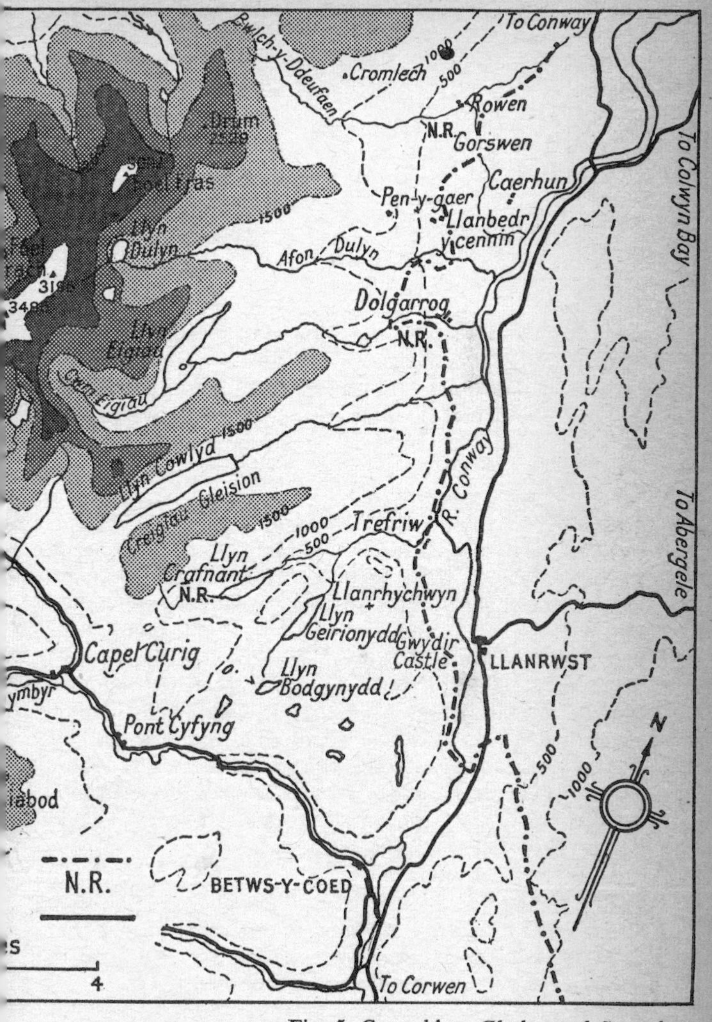

Fig. 5. Carneddau, Glyder and Snowdon

squared slabs and columns that it is no wonder that people in the past have considered them the ruin of something artificial, some shattered mountain-top Stonehenge. It was Lhuyd who first described the top of Glyder Fach for the world, in the *Britannia* of 1695. Since Lhuyd many others have both described it and argued about its origins. Again, the Reverend John Evans is quotable here if only because he shows how people were thinking about the earth in the eighteenth century. Evans was a reactionary. He could no more accept the volcanic theory that some were beginning to propound to explain the origin of some of the Snowdonian rocks than he could accept the theory of evolution towards which Buffon had been groping his way a few decades earlier. Evans would have been unimpressed by the theory of glaciation had he lived to hear it. For he had perfect faith in the action of water as an explanation of most geological phenomena, a faith that rested on his unshakable belief in Noah's flood. So Pennant's remarks on the summit stones of Glyder Fach were unacceptable to Evans. Pennant had said, obviously with the new volcanic theories in mind: 'I should consider this mountain to have been a sort of wreck of nature, formed and flung up by some mighty internal convulsion.' But, asks Evans astutely, what about the fossils? Surely, since many of these rocks have 'marine shells imbedded in them, they rather support the opinion of the action of water than the action of fire'. But what neither Lhuyd nor Pennant nor Evans nor that other amateur geologist Aikin (who also rejected the volcanic theory) ever realised was that the volcanic rocks of Snowdonia are interstratified with fossil-bearing sedimentary rocks, a fact not demonstrated until the early nineteenth century.

A summer's evening is the perfect occasion for seeing these fragments of a shattered world that lie grouped along the Glyders, or for clambering among the jagged summit rocks of Tryfan. For at that time of day, with the sun low in the sky, not only are the shadows of the columnar rocks very beautiful but you can also see more clearly the strangely quadrangular shapes of these columns that made the earlier writers think of Stonehenge but which are in fact due to the natural splitting or jointing of rocks that have been subjected to great pressure or heat. Symmetrical blocks frequently result from such jointing and the faces of the joints are often smoother and straighter than the surfaces of true strata. Such faces provide climbers with some of their most interesting problems.

This last hour of evening sunshine, when the tops are lit with a wonderful light, is also the best time to look from a high place across the rest of the mountains and see details of their shapes that are lost in the light of midday. As the setting sun's light falls along the slopes you become aware of countless knolls and hollows and insignificant ridges you may never have noticed before. And the glaciated smoothness of the valleys then becomes particularly clear. Nant Ffrancon from Glyder Fach looks deeper and more rounded than ever; and the stream is a silver wriggling snake in the blackness of the shadow-filled valley.

It is a long and not wholly pleasant scree that takes you down from the Tryfan-Glyder Fach saddle into the depths of Cwm Bochlwyd. There are few plants on it except the ever-faithful starry saxifrage; but ring ouzels pipe from the crags or dash chattering angrily down the slopes if you chance to be near their young. And wrens, as in every Snowdonian cwm, sing piercingly or stutter excitedly at you from the shelter of scattered boulders. In Bochlwyd you are nearing sacred ground again, I mean the rocks that contain the calcareous ashes that provide a habitat acceptable to the rarer alpines. Llyn Bochlwyd reminds me how times have changed and how much the mountains are now invaded. Ogwen and district seemed very popular before the Second World War but in those days Cwm Bochlwyd was still reckoned pretty remote and was not much visited. Today you will probably see plenty of climbers' tents there, grouped among the grey rocks round the lake like the hut-circles of Bronze Age man. It is a strange thought that the campers of our time are possibly the first people to live, however temporarily, in Cwm Bochlwyd since the days of Bronze Age man. Not that there is any indication as far as I know that Bronze Age people ever did live there. But it would be strange if they did not. For Cwm Bochlwyd with its deep clean lake and its mighty encircling cliffs is one of the truly magnificent retreats of Snowdonia.

When it comes to translating the word Eryri, which is the old Welsh name for the Caernarvonshire mountains, you can choose between Camden (1586) and Lhuyd (1695). Camden, firm in his belief that these mountains lay white with snow throughout the year, had no difficulty in connecting Eryri with the Welsh word *eira*, which means snow: 'It harbours snow continually, being throughout the year covered with it, or rather with a hardened crust of snow of many years continuance. And hence the British name of Creigiau Eryri, and that of Snowdon in English.' But Lhuyd was equally definite: 'The British name of these mountains, Creigiau Eryri, signifies Eagle Rocks, which are generally understood by the inhabitants to be so called from the eagles that formerly bred here too plentifully.' But having said that he wavers a little and concedes that there might perhaps be something in Camden's snowy mountain idea: 'Seeing the English call it Snowdon, the former derivation was not without good grounds.' The word Snowdon, whatever its derivation, is also old, for, says Pennant, it is found in Saxon documents which spelt it 'Snawdune'. As for 'Snowdonia', this is a name that has had some quite unjustifiable insults thrown at it. It has been called 'a vulgarism' and been blamed on Pennant. Yet it has in fact a respectable antiquity and origin. 'Snowdon' used to mean not one mountain as it does today but the whole of upland Caernarvonshire. So naturally, in the days when scholars wrote in Latin, Snowdon, or rather Snaudon, was latinised into 'Snowdonia'. Camden reported in 1586 that 'Snaudonia' had long been in use by historians.

Lhuyd summed up very clearly the chief characteristics of the Snowdonian heights: 'Such as have not seen mountains of this kind are not able to frame an idea of them, from the hills of more champain or lower countries. For whereas such hills are but single heights or storeys, these are heaped upon one another, so that having climbed up one rock, we come to a valley, and most commonly to a lake; and passing by that, we ascend another, and sometimes a third and a fourth, before we arrive at the highest peaks. These mountains, as well as Kader Idris and some others in Meirionydhshire, differ from those by Brecknock and elsewhere in South Wales, in that they abound much more with naked and inaccessible rocks; and that their lower skirts and valleys are always either covered or scatter'd over with fragments of rocks of all magnitudes, most of which I presume to have fallen from the impendent cliffs.'

In medieval times there was a royal hunting forest in Snowdonia. According to Leland, it lay wholly in Caernarvonshire, not extending into Merioneth: 'All Cregeeryi [Creigiau Eryri] is forest and no part of Merionethshire lieth in Cregeeryri.' In this statement Leland almost seems to foresee the controversy that was to rage about this very point not many years later. 'Forest', in the sense he used it, meant not a tree-covered place but rather the opposite: a mountainous place where presumably there would be good hunting of deer, boar, fox and other animals driven up from the trees on to open ground. Pennant tells us that 'Snowdon being a royal forest, warrants were issued for the killing of deer.' He had himself seen three such warrants dating from the sixteenth century. One of them, from Elizabeth Tudor's reign, 'was addressed to the master of the game, ranger and keeper of the queen's highness' forest of Snowdon in the county of Caernarvon'. Elizabeth, says Pennant, tried to gratify the rapacity of her favourite, the Earl of Leicester, not only by giving him the rangership of Snowdon Forest but also by trying to extend the forest to include both Merioneth and Anglesey. The fact that a Snowdon deer was chased across Menai Strait and killed at Malltraeth several miles away was accepted (by a packed jury) as proof that Anglesey was part of Snowdon Forest! Leicester would have got away with this sharp practice had not his annexations understandably roused the indignation of an Anglesey landowner, Sir Richard Bulkeley of Baron Hill, Beaumaris. The fact that Bulkeley was also one of the Queen's favourites enabled him eventually to defeat Leicester's schemes. Lewis, the topographer, states that the royal forest of Snowdon 'was erected by Edward I on his entire subjugation of the Welsh' and that 'it was finally disafforested about the year 1624, to the great satisfaction of the neighbouring farmers whose crops suffered greatly from the deer'.

From north, east, south and west there are well used tracks to the top of Snowdon. To the earliest tourists, in those late eighteenth-century days when there was no road up Llanberis Pass but there was one from Caernarvon to Beddgelert, it was the Snowdon Ranger track that was best known. The route from Llanberis was also used, but in those days to get to Llanberis from the coast involved going by boat up the lake. Three routes were recommended in the 1790's. First: from the west by the Snowdon Ranger track. Second: from the north up what is now the railway track. Third: from the south-west near Beddgelert ('but this is extremely steep and dangerous'). It was not until the road was made up Llanberis Pass about 1830 that the now so popular eastern approach began to get known to many people. This fourth approach, by way either of the Pig Track or Crib Goch, is generally thought of as the finest way up Snowdon, though the fifth way, the Watkin path, is almost as good.

The name Pig Track is quaintly altered to 'Pyg Track' on the current 1-inch and 2½-inch Ordnance maps, which is regrettable because the present spelling masks the fact that this route was named Pig Track because it passes over Bwlch Moch—the Pass of the Swine, and so is probably a memory of the hunting of wild boar in medieval times. Bwlch Moch, whose name may be in danger of being forgotten for it is not on most maps, is the gap the Pig Track passes through a mile from Pen-y-pass. It gives you that first splendid view of the slopes and precipices that stand round the Snowdon Horseshoe with Llyn Llydaw in the centre.

There are other routes but they are mostly precipitous and are only for the climber, and sometimes the botanist, for their ways not infrequently cross. When they do there is co-existence rather than fraternisation. For climbers have commonly in their jargon the word 'gardening', which means tearing the vegetation off the rocks to make them more climbable. It is an activity that botanists regard with marked coolness.

If I were asked where to start by anyone who wanted an introduction to the natural history of Snowdon I would recommend him to begin down at Llyn Gwynant. By this

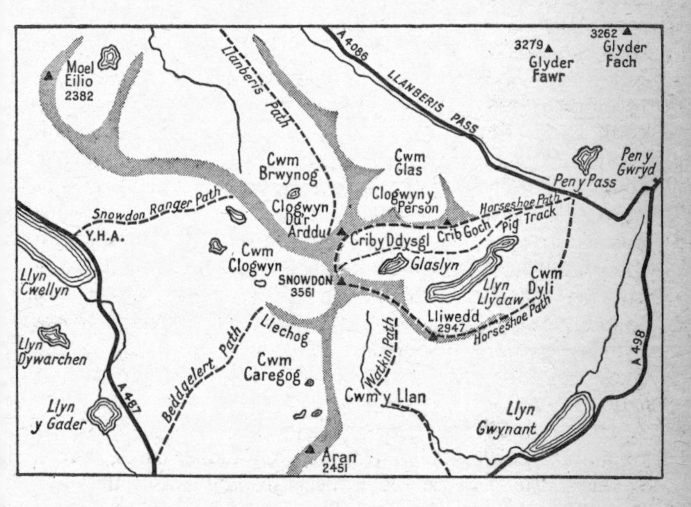

Fig. 6. The summit ridges of Snowdon

most beautiful lake the abundance of ash trees immediately suggests a lime-rich soil and this is confirmed when on the tree-covered rocks above the lake you find the graceful rock stonecrop (*Sedum forsteranum*), a plant strangely rare outside Wales and the south-west of England, except as a gardener's escape. Going up the valley towards Pen-y-gwryd you soon lose these lime-rich rocks for acid ground but you find them again higher up, in side-dingles such as Ceunant Mawr where there is quite a wealth of semi-upland species characteristic of the shady, wet and less acid rocks. From Ceunant Mawr you can cross the valley and ascend Cwm Dyli to see the first good cliffs for alpine plants rising on your left as you approach Llyn Llydaw, cliffs which are far richer in species than the main cliffs of Lliwedd are. Then across the lake by the causeway and up the rocky slopes of Crib Goch to the Pig Track, noting the spectacular sheets of sloping glaciated rocks alongside the Pig Track in many places.

No plant-seeker will need telling that at any rate this part of the Pig Track crosses rather acid ground for the rocks are poor in species. But if, when between Llydaw and Glaslyn, you leave the track and go directly up the slope, you will soon come within sight of the dip in the skyline called Bwlch Coch, which offers a neat way of slipping round the fangs of Crib Goch. Look now, before coming up to the gap, at the rocks about you and you will see quantities of moss campion, purple saxifrage, vernal sandwort, rose-root, alpine meadow-rue, alpine scurvy-grass, mountain sorrel and mossy saxifrage. In other words you have come to a typical group of mountain plants, mostly calcicoles, flourishing reasonably well although they are on south-facing rocks and are therefore exposed to the drying effects of the sun and the prevailing wind. But here, though they do at times have to face droughts, they are high enough to enjoy Snowdon's special climate, the mists, the many sunless days, the heavy rainfall and snowfall, and there is always at least a seepage of water to their roots. Not far above, you come to the top of the ridge past the Nature Conservancy's squares of fenced land that are being used for high-level grassland studies (about 2,500 ft.). In the col here, which is on the Snowdon Horseshoe track, you will, especially at holiday-times, see a spectacle not encountered on many mountains: the herring gulls which hang very tamely about parties of mountaineers for scraps of food, for this col is a popular spot for a rest and a bite to eat. In the eighteenth century probably the only gulls hereabouts were the great black-backs that then nested on an island in Llyn Llydaw and, says Pennant, 'broke the silence of this sequestered place by their deep screams'. They have gone long since, but for years now, and in increasing numbers, the herring gulls have frequented the Snowdon Horseshoe to scrounge round picnicking parties. Similarly, but only very recently, the black-headed gulls have learned that they have only to appear at lay-bys along Snowdonia's main roads and people will almost shower food upon them. While at this gap of Bwlch Coch turn your binoculars on the summit of Snowdon and on a fine holiday you will find it hard to decide which is greater, the number of people on top or the number of herring gulls. While you are looking at the peak look also at the cliffs immediately under it, which are called Clogwyn y Garnedd, and see how beautifully the strata lie there like a pile of deep saucers one upon the other, each saucer smaller than the one below it, showing very clearly that the summit of Snowdon, though it reaches to 3,560 feet, is only what time and the forces of nature have left of the *bottom* of an ancient *down-folding* of the earth's crust.

From these great east-facing cliffs come many springs and streamlets which unite lower down into a waterfall that on still days is the dominant sound of the head of this fine cwm, a sound sometimes mingled with the laughter of herring gulls, the distant voices of people moving along the higher and lower tracks and—cruel shock to the unsuspecting lover of mountain wilds—the heavy chuff-chuff of the train on the last steep pitch to the summit.

From the high gap of Bwlch Coch you can clamber to the right up the bristling pinnacles of Crib Goch, or to the left up the ramparts of Crib y Ddysgl and so to the top of Snowdon, called in Welsh, Y Wyddfa, which probably means the Tumulus. But if so there is nothing known to archæologists about it and it should be classed with Cader Idris as a mountain-top name invented to adorn a legend. With thousands of people going up Snowdon every year it is inevitable that the original vegetation has long been trampled out of existence, and any seeds of alien plants accidentally carried up, though they might germinate, are unlikely to survive the winter. It is rare that we can be definite about the height to which any species of small mammal ranges but we can certainly claim the house-mice observed at the summit hotel as the highest in England and Wales! Apart from the herring gulls, birds are naturally few on this over-populated summit. I have seen a couple of lesser black-backed gulls and a wheatear and there are often meadow pipits and ravens not far away, and ring ouzels in the crags just under the summit. But though Snowdon cannot claim to be a bird-watchers' Mecca, it is saved from ignominy by its choughs. In chapter 5 I mentioned seeing choughs high over Lliwedd in late summer. At that time of year I have also seen them down in all four of the chief valleys under Snowdon: Cwm Dyli, Cwm y Llan, Cwm Clogwyn and Cwm Glas. But they are far from common round Snowdon. I have seen only pairs and family parties of five or six birds, never a flock, though presumably there are flocks at times for choughs are very gregarious, especially in autumn, and evidently gather from considerable distances. As these Snowdon choughs seem always to nest in quarries or mines and never in natural rock-clefts, can we assume that they did not start breeding on Snowdon until man came and quarried there? Choughs are grassland feeders. They spend hours walking quietly about the slopes and near the lake-edges pecking into turf and every now and then shifting their positions by flying thirty or forty yards to try somewhere else. When they fly they usually call to each other, otherwise they would seldom be noticed. What they eat I have never been able to see except that it is nearly always something very small, presumably a variety of small creatures of the turf such as ants and spiders. In the hard weather of 1962–3 numbers of choughs were driven down from the mountains by the snow and one or two were even reported feeding in gardens on the coast. Two were brought to me that had died of starvation and it seems probable that this beautiful species is not altogether hardy. This would explain why it is limited in Britain to the west, presumably holding out in mountain districts only where it can enjoy the shelter of deep quarries.

A word about the famous eagles of Snowdon. We can reasonably assume, if we cannot absolutely prove it, that until a few centuries ago eagles were widespread in the mountains of England and Wales just as they are in Scotland even today, for there are so many references to them in literature, tradition and place-names. But there has been no occasion, as far as I know, of anyone in Wales at any time ever stating categorically that he himself had seen eagles breeding at any particular place. This is not so surprising as it may at first appear if eagles had virtually ceased to breed before the end of the seventeenth century, for it was not until then that people first generally began to see any point in recording natural phenomena of this sort. It is true that the seventeenth-century naturalists Johnson, Ray and Willughby mention eagles in Snowdonia but without claiming that they themselves had seen one. Then a little later Lhuyd, as I have mentioned, reported eagles at the end of the seventeenth century as having been too common in the past. He implies that it was the recent past, for he goes on to say that 'they do yet haunt these rocks in some years, though not above three or four at a time, and that, commonly, one summer in five or six; coming hither, as is supposed, out of Ireland.' How tantalisingly vague this is! If only he had given us just one brief mention of an eagle seen by his own eyes. As it is, I fear his words sound like those of one who had heard much talk of eagles from the local inhabitants but never saw one there himself. The picture that emerges from all the accounts seems to be that golden eagles bred regularly in Snowdonia until about the middle of the seventeenth century, that they were then much persecuted by farmers and shepherds, and had probably ceased to breed about the year 1700 or soon after. This did not prevent writers throughout the following century speaking of eagles being still present. Pennant mentions them (but did he ever see one?) and so does Pennant's parrot, the Reverend Evans, who adds this inevitable picturesque touch: 'These carnivorous birds of prey in many places formed a formidable banditti; seizing upon the poultry, sheep and even young children.' Such beliefs endure. I am sure that if eagles were to try to nest in Wales today we would hear an outcry of similar nonsense.

Botanists since the earliest days on Snowdon have concentrated on Clogwyn Du'r Arddu, Cwm Glas and Glogwyn y Garnedd. Of these Cwm Glas is the most magnificent and the richest in flowers. Shut off on three sides by mountain walls and on the fourth by a splendid drop into Llanberis Pass it is less overrun than the rest of Snowdon is. Floristically, owing to the depredations of generations of collectors, especially the conscience-less fern-enthusiasts of last century, Cwm Glas is no longer quite the rich locality it was in the times of Johnson, Ray and Lhuyd and you will not find the holly fern there a tenth as easily as they probably could. Perhaps the Pig Track from Pen-y-pass and then the climb over Bwlch Coch is the best way into Cwm Glas. The Crib Goch ridge itself is barren enough but you are not many yards down towards the lake before you become aware of the increasing richness of the vegetation. Not that the plants here will stagger you with their variety. You have already seen most of them as you came up from Llyn Llydaw. What is impressive is their health, vigour, abundance and green sappiness; the extra-large specimens of lady's mantle, rose-root, wood-rush, moss campion, mountain sorrel, alpine saw-wort, thrift, scurvy-grass, alpine meadow-rue and vernal sand-wort, filling every space and corner of the damper ledges and gullies. In shady crevices there are quantities of the commoner ferns: green spleenwort, bladder fern and beech fern. And very beautiful it is to see the plants of lower ground flourishing up here too: globe flowers, ox-eye daisies, water avens, stone bramble, golden-rod and, making the brightest patches of colour in the whole face of the cliffs, the common but splendid red campion. It was in a pool in this cwm that Lhuyd made the first British record of that very small submerged plant, the awlwort. The pool lies about 300 yards from Llyn Glas and at a higher level. Nameless on the 1-inch map, it is Llyn Bach, the Little Pool, on the 2½-inch map. In Lhuyd's day it was Ffynnon Frech, which means the same thing. It is fascinating to find this little pool among the rocks and to see still growing in it the same plants that Lhuyd saw there two and a half centuries ago. Surely, in a world cursed with the constant despoliation of natural habitats, the long duration of all these mountain plants is something for which to be thankful.

CHAPTER EIGHT

Hebog, Moelwyn and Siabod

Because one of the best week's mountain walking I have ever enjoyed was spent on these three delectable ranges of Hebog, Moelwyn and Siabod during a time of perfect September weather, I can think of no better way of introducing this part of the Park than to describe where I went that week and what I saw.

I set out up Cwm Pennant, the valley a few miles north of Criccieth, long celebrated by the bards for its beauty. Its clear, alder-shaded stream, the Dwyfor, comes winding down the lower half of the cwm through pastures and many scattered oaks and ashes. So much ash there indicates the good soil that often occupies the bottoms of valleys where drainage has washed the mineral wealth off the hillsides into the bottoms. There are two distinct habitats, one rare, one common, where ash flourishes best in Snowdonia. The rare type is a dry limestone mountainside such as at Craig y Benglog near Dolgellau. The common habitat is well-watered valley floors and riversides. Often a pale band down the face of an oak-wood shows where ash trees are growing along some stream that comes down the woodside there. So pronounced is this difference between the two habitats of the ash that some botanists have drawn a distinction between 'limestone ash' and 'water ash'.

Inevitably fertility diminishes up the Pennant valley. Bare rock soon begins, purplish outcrops and grey glacial boulders among tall bracken, and the scene is immediately wilder. You cross the stream at Pont Gyfyng—the Narrow Bridge—go up a little hill and round a bend, and the great grassy slopes of Moel Hebog are before you, divided into immense 'fields' by walls reaching right to the top of this shapely, and on this side cragless, mountain. From here, too, you see where the valley reaches its end, or rather its beginning, in the striking half-circle of hills, some smooth, some precipitous, that goes from Hebog to Moel Lefn and round the head of the valley to the fine hills that look away on the other side down the crags of Cwmsilyn into the Nantlle valley. Up the road, which follows the stream that is now only three yards wide and full of black, darting troutlets, I overtook an old man who reminisced about days half a century ago when, he told me, Hebog was a popular mountain and people came in crowds, sometimes in brakes but usually walking all the way from Criccieth to the top and back, which, he said, 'was the best part of twenty miles, some of it very rough. But people come in cars now and hardly anybody can walk any more.' Incidentally, the map-name, Moel Lefn, was unknown to him, his name for both that hill and the one between Lefn and Hebog being taken from the farms below them.

It was already late afternoon when, two miles north of Pont Gyfyng and near the valley's last farm, I crossed the Dwyfor and set off up a side-stream whose banks were gay with devilsbit scabious and sneezewort. I climbed the lower slopes of Moel Lefn up through little rough fields enclosed by walls. At first there were willows along the stream and isolated thorns in the pastures. Then the last tree was behind me except for a few final rowans, splashed scarlet all over by a wonderful crop of berries. On the open hillside a green woodpecker rose from an ant-hill and loudly called *plee-plee-plee* when it landed on top of an oak far down the slope. Wall butterflies sipped at the yellow disks of the autumn hawkbits. I passed up through a short zone of bracken into a low-growing patch of gorse now bright with yellow flowers. Soon the ground became boggy where springs came out of the hill and there was no longer gorse or bracken, only short turf and sedges, lesser clubmoss and sundews: the sort of place, if it had been a little more lime-rich, to look for grass of Parnassus, for September is one of its best months. Not that it would have survived here in this close-nibbled turf. For all about grows the mat-grass whose wiry stems the sheep dislike, so they concentrate their attention on the better variety of plants in the wet places round the springs. If you look into the turf you see that the stems of countless plants which have aspired to grow upward have been nipped off. Such species must vegetate here indefinitely without being allowed to form flowers or fruit. Perhaps the bog orchid is not quite the extreme rarity we think it is: perhaps it is simply that the sheep hardly ever allow it to develop. The plants that suffer least in such sheep-infested circumstances are those such as sundews and butterworts whose leaves lie flat and whose flowering stems are usually too insignificant even for the sheep to bother with. Butterworts probably have the added protection of being distasteful to sheep.

I went through a gate in the mountain wall—the long contouring wall at about 1,500 feet that separates the highest sheepwalks from the lower ground—and now the ground got rockier, the alpine clubmoss was crisp under my feet, and with delight I heard the vehement, sneezing cry of a chough: *chee-ow, chee-ow*. Looking up the mountainside through my binoculars, I watched a curious sort of display-combat between two choughs on the highest screes of Moel Lefn. They 'fought' with a flurry of black wings and loud cries for a few minutes, then broke off abruptly to peck quickly about in the grass (many brown ants were scurrying in the turf that sunny afternoon). Soon they both rose and came flying down the hill, the one diving at the other in a series of quick swoops that were beautiful to see, for few birds are more acrobatic on the wing than choughs. As they passed me they tilted away on curving flight, the bright sunlight catching the brilliant red of their legs. Then they vanished round the slope to the north.

Soon I came to the top of Moel Lefn, a mass of bare rock surrounded by shattered fragments that reminded me of the summit of Aran Fawddwy. Meadow pipits flitted about the stones—how they love the mountain tops in calm, warm weather!—but I looked in vain for either the stiff sedge or the least willow that are so often near each other around summit rocks. But I suppose Moel Lefn, only just over 2,000 feet, is not high enough for these true mountain plants. On a clear day the view from Moel Lefn is a spacious one. But you have to pay for your sunny anticyclones by doing without far prospects, and that day Moelwyn was invisible under a grey blanket that lay in a straight line all along the southern horizon. Even Moel Hebog only a mile and a half away loomed vast and remote as it vanished and reappeared as the haze thickened and thinned. To look down the east side into the forestry plantations that fill the entire cwm there was like looking into a huge black lake. The only clear thing in the landscape was Snowdon itself, for its uppermost three or four hundred feet stood sunlit above the murk. The highest part of the well-worn Snowdon Ranger track made a prominent reddish streak below the summit.

Southwards from Moel Lefn you dip and rise again to a similar rocky top, with fine slabs of rock much jointed and split into columns. Some of the rock here is as rough as a very coarse conglomerate, the surfaces being armed with sharp protuberances. Few plants, of course, grow amid such poverty and exposure. There was the inevitable woolly-haired moss in grey clumps here and there. There were fir clubmoss, bilberry and green lichens. On the sheltered east side were a few small leathery patches of cowberry, its pinky-white bells only now just opening (15 September). Then everything changes dramatically as you come down into the narrow pass under the north slope of Hebog. You notice that green lichens have mostly ended and that a thickly encrusting white lichen has replaced them; and that now instead of hard grey bare rocks there are brown, softer-looking rocks with many holes and ledges and fissures with green plants bristling out of them, not bilberry now nor cowberry, but rose-root, golden-rod, beech fern and devilsbit scabious. And as you follow these rocks round, the plants get choicer. Here is a little mountain everlasting, leaves pale-green above, silvery-hairy below. Next to it is the only rose of these mountain ledges, the burnet rose with its multitude of fine sharp thorns. Eventually you scramble down into a wide gully where

Nantlle
To Caernarvon
Y-Garn
2080
Trum-y
Ddysgl
SNO
Carnedd
Goch
2220
2408
Llyn
Gadair
1500
Yr Aran
2451
Llyn
Llywelyn
Afon Dwyfor
Moel
Lefn
2094
Cwm
Meillionen
1000
Bwlch
Meillionen
Moel Hebog
2566
Beddgelert
500
Pont Gyfyng
1500
Aberglaslyn
500
Llyn Cwm-
ystradlyn
Pen-y-Gaer
Moel Ddu
1811
Nantmor
1000
Croesor
500
CRICCIETH
N.R.
Tremadoc
Llanfrothen
PORTMADOC
Afon Dwyryd
Traeth Bach
Morfa Harlech
To Harlech

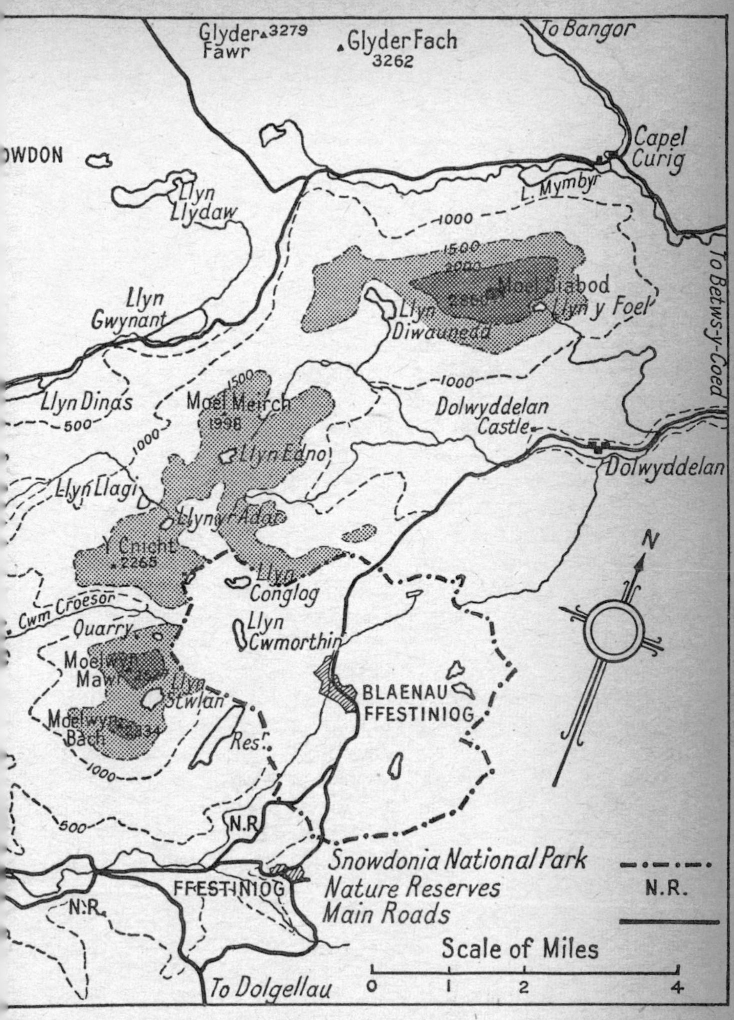

Fig. 7. Hebog, Moelwyn and Siabod

the rocks are black and dripping and the ledges are crammed with more rose-root and prosperous plants of mountain sorrel and lesser meadow-rue. There are starry and mossy saxifrages everywhere and a rich variety of mosses and ferns, including green spleenwort and bladder fern.

Under one of the walls of this gully a small cavern was hollowed out about the end of last century as a trial-hole for asbestos. This curious substance lies in a band two or three inches thick round the walls of the cave in close company with a narrow band of quartz. It is white, wet and soft—you can claw it out with your fingers—and much resembles a scentless sort of putty. It contains fibres rather like those of wood from the heart of a rotten tree but they are too short to be of industrial value and the trial-hole was soon abandoned. No road was ever made to it, the asbestos being humped down on men's backs or on ponies. Following a shelf round this excavation I came to the natural cavern—a mere overhang of the rock—called Ogof Owain Glyndwr, from which this hill between Hebog and Lefn is named Moel Ogof.

Not far below the cave begin the conifers, mostly Norway spruce, planted in the mid-thirties and growing well, as so they should for this cwm is Meillionen—the Valley of Clover, a name suggestive of good land. Remembering that it was during the early days of this forest (about 1937) that two pine-martens were unintentionally trapped here, I looked hopefully about me as I made my way up the scree of Moel Hebog. But I saw no animals—only plants: a juniper bush, the three common clubmosses and those usual scree-dwellers, bilberry, woolly-haired moss and parsley fern. On my left were glacier-polished rocks and small cliffs hollowed under by what looked like more trial-holes for asbestos, or possibly gold, for there is much quartz in the rocks just there. The surprise of Hebog is that despite all the scree on the way up you find hardly any rock at the summit. Instead there is a narrow grassy ridge that looks down steep craggy slopes to Aberglaslyn Pass on the east and Cwm Ystradllyn on the south.

The sun had gone down into a bank of cloud between

Lleyn and Anglesey and it was already dusking before I reached the top of Hebog. The lights of houses were showing in Beddgelert but there was still enough daylight to see Llyn Dinas and Siabod standing high and dim beyond. Snowdon, Crib Goch and Lliwedd were still clear, and in front of them Aran, sharper-pointed even than Snowdon. In the valley to the north Llyn y Gadair was a pale gleam in the shadows and I could just make out Llyn Dywarchen and the great rock that stands up in it. I thought of Giraldus and his party going along the valley in 1188 and turning aside to see the wonder of Dywarchen's floating island, which was a large slab of peat that had detached itself from the bottom of the lake. I thought of the astronomer Halley swimming out to this unsubstantial island in 1698 to satisfy himself that it really did float. And I thought of Pennant and all his successors dutifully making their way there too in the following centuries, to see this by then diminishing marvel which has now quite gone. Turning south I watched Cader Idris, then Rhinog, disappear into the night. But now Moelwyn and Cnicht had cleared and were catching a last pale light, enough for me to see something of the way I intended to go next day.

All round the west and north the sky was aflame with the setting sun. Holyhead stood up in blackness and there was peak after little peak all along Lleyn, only their tips visible, poking out of the mist. But Cardigan Bay was clearer: I could see the lights of the houses of Abersoch; the little St. Tudwal's Islands, one beyond the other; and, like a full stop at the end of Lleyn, the hump of Bardsey Island. I thought of the bird-watchers in the observatory there and hoped they were recording more birds than I was seeing on Hebog. Bardsey light is not visible from Hebog but South Stack and Skerries were flashing in the north-west off Anglesey. Beyond them, beyond all Anglesey and Lleyn, and beyond the sea, lay many undulating miles of the Irish mountains cut out black under a bright crimson belt of sky. I sat so long in the warm dusk at the top of Hebog that it was already dark and starlit as I made my way down the west side; yet it is surprising how much light the pale grass holds and all the way down I could see the difference between dark rushes, light grass and palely gleaming rocks. I unrolled my sleeping-bag by a murmuring stream and went to sleep looking at Jupiter bright over Hebog and thinking of the botanist J. Lloyd Williams who, when a young schoolmaster here years ago, found the Killarney fern, Snowdonia's rarest species, along one of the streams on this side of Hebog. It has not been seen since because the precise locality was never recorded; but it probably grows there still in the spray and shade behind some little waterfall. So there is a challenge!

Early next morning I made my way by road to Llanfrothen, for the direct route into the heart of Moelwyn is the road up Cwm Croesor under the steep south side of Cnicht.

It is a rough road but a straight one that takes you boldly up the valleyside to Croesor quarry. The road is still in use by lorries because although no slate has been worked here since about the 1930's, the mine's spacious underground cathedrals are thought a safe place for storing explosives. I arrived in time to see a little engine and a train of wagons disappear into the mountain, its muffled sound getting less and less until I could hear nothing. I looked at the ferns that have made free use of the lime-mortared walls of the quarry buildings here. There were ten species in a few yards either in the wall or on the bank below it: bladder fern, polypody, hartstongue, common spleenwort, parsley fern, male fern, broad buckler, hard fern, mountain fern and lady fern, and I expect wall rue and rusty back could have been found not far away. High above the quarry the usually dark north-east-looking cliffs of Moelwyn Mawr were bright in the morning sun, but in an hour they would be back in the shadows which they and their plants experience most of the time. The gullies up there have been gouged out of fairly lime-rich rock and have some interesting calcicole plants. But as I had botanised there a few months before, I now turned my back on Moelwyn and followed the small cliffs round the head of Cwm Croesor. I have never had time to search these cliffs, always being bound elsewhere, but they bristle with rose-root; and as rose-root is often the first step to something good, there may be quite a variety of species on those ledges.

From there it is only a few hundred yards across wet moorland and ice-smoothed rocks to the desolation of abandoned quarries at the head of Cwmorthin. This was a very lonely and beautiful place in the eighteenth century. Out of all the wonderful wild places Pennant saw in his travels, he singled out Cwmorthin as especially sequestered. The nineteenth century changed all that when man made his violent intrusion. For many years the slate-veins were followed deep inside Moelwyn Mawr and Moel yr Hydd. Enormous tips of slate-waste were spewed out onto the slopes below the mine-levels. Everywhere were men, railways, engines, buildings, dams, reservoirs and turning machinery.

Now it is all long finished. I looked down to where, below and beyond the tips, long roofless buildings and shattered single walls stood silent in the soft autumn sunlight. I walked down to the ruins. It was like discovering the remains of a lost civilisation. My shoes clattered loudly on broken slate. Two choughs rose from a building and scolded me from the hillside with high, churring, nasal cries, their black feathers glossy in the bright light. I peered into the dripping darkness of a level-entrance, standing awhile in the cold draught from the mine, for slate-quarries are dismally cold inside. On the ground among the broken slates grew parsley fern, quantities of the tiny New Zealand willow-herb and, inevitable relic of man's occupation, stinging nettles. I clattered out of the ruins along more enormous tips to look down into Cwmorthin, a colossal gash of a valley with scars and cliffs high up both sides. I could see two abandoned houses, a derelict chapel, a desolate lake and the empty quarry road winding out of sight down towards Tanygrisiau. And I was alone in all that world of mountain, valley and past human labour; myself, two choughs, a handful of shrill-voiced pipits and the quiet sheep of the hills.

I turned north past a bog-filled hollow that had been a

reservoir until its dam decayed. I scrambled up small cliffs, following a waterfall; then across rocky and grassy hills to a strange, rarely visited region of moorland and bog and a bewilderment of crags and natural lakes and quarry reservoirs. It was all acid ground. 'Hopeless sterility' was the damning verdict on this district by one of the early botanists, and most others seem to have followed his lead, for this seems to be botanically the least known section of the Park, especially the part lying in Merioneth. Most of the rocks here are dolerite which sometimes produces mineral-rich soils, but here it seems to be very lime-deficient, judging by the green, acid-loving lichens that colour the rocks and the absence of any calcicoles. Still, acid or basic, all the crags in this area would be worth searching. There is an old record of forked spleenwort, and a still older one of northern rock-cress. There are rocks near Llyn Llagi that are distinctly floriferous, and all the countless little bogs are worth looking at for the rarer bog mosses, sedges, long-leaved sundew and—what would be a great prize—the marsh clubmoss which was found years ago at some now unknown locality in Merioneth.

Two green sandpipers rose steeply from the edge of a small reservoir. I watched them for several minutes as they wandered about the sky, at first together then widely separate, then together again, their shrill *skittly-wit, skittly wit-wit-wit* coming back from far across the moor. Patterns of long, pale-green fronds of the flote-grass lay in curving lines on the still surface of the water. I ate my lunch above a nameless little natural lake deep in the rocks. The sun lay warm about me and there was an indistinct view down Cwm Croesor to Portmadoc, Borth-y-gest and, across the estuary, to the shining sandhills of Morfa Harlech. A large yellow-and-black dragon-fly flew back and forth near to me, making lightning dives on to flies at the water's edge. A silver-y moth, the first of the year, hovered over heather flowers. The only butterfly I saw (the small heaths were now over) was a large white, fluttering steadily past against the light breeze. When I reached the next ridge all the mountains of the north rose up, looking Himalayan through the haze. The south side of Snowdon was now very close.

Siabod stood up in isolated splendour as Siabod does from every angle. But an even thicker haze which had lain all morning as a deep purple band on the sea in Cardigan Bay was beginning to invade the hills. Hebog was vanishing and Snowdon would not be clear much longer. Already it was hopelessly unphotographable. Glorious anticyclones are excellent for mountain walking but are death to landscape photography. All round you the sun is brilliant on every rock and blade of rush. But distance is annihilated.

I walked across an autumn-red cotton-grass moor past blue-black peat pools edged with brilliantly green bog mosses. Llyn yr Adar, the Lake of the Birds, presumably named from some ancient colony of black-headed gulls, was a pale-blue circular pool down on my left. For a moment I was tempted to follow its outlet stream down the rocks to Llyn Llagi, where I could have seen some good plants such as green spleenwort and northern bedstraw.

Another mile north brought me to Llyn Edno, a perfect little lake of attractive, irregular shape and very blue that day among the pale-green grass of its slopes and the even paler rocks. Immediately above the lake rose the jagged ridge of Moel Meirch outlined not against the sky but against the grey bulk of Glyder Fawr, five miles beyond. Another white butterfly passed migrating directly south across the hills. For several minutes I followed it through my binoculars until it was through a gap in the rocks. It never glided as some butterflies do, which you would think might give the wings a rest, but went fluttering on and on following its steady line. How do such fragile creatures store up the energy required to work their wing-muscles on journeys that may be hundreds of miles long?

The view of Snowdon from Moel Meirch is directly up Cwm y Llan. Through the deepening murk I could just make out the line of the old quarry track leading up to the Watkin path. A little nearer and clearer was the south flank of Lliwedd with its disused levels and mine-tips one below the other, following the ore in a straight line down the slope. I remembered the plant-rich rocks a little to the east of those mines and contrasted them with the arid slabs and columns all about me. But even the acid heathery rocks of Moel Meirch have their plant of distinction—the juniper, here draping itself over the ledges with unusual luxuriance, making flat bushes five or six feet across. This is the mountain juniper, so very different in appearance and habitat from the juniper common on the chalk of England. The chalk juniper is usually an upright shrub with prickly needles that stick out at right-angles to their stems; and it is an out-and-out calcicole. But the mountain juniper, regarded as a distinct species by some botanists, is a shrub of acid rocks, lying prostrate across them and having its needles pressed to the twigs and unprickly. The difficulty of making this alpine juniper into a separate species is that so many intermediate forms occur that the two extreme types appear to run into one another. The same difficulty arises when you try to separate the mountain and lowland forms of golden-rod.

I shared the warm summit rocks of Moel Meirch with four large hover-flies that kept flying from rock to rock with a noisy buzzing of wings. When they settled they produced their familiar high-pitched singing whine which I suppose is some sort of mating call, for every few minutes they made darting sallies at each other. I noticed that each hover-fly was silent when it first settled and that the singing did not begin until, by a quick movement, the wings closed tightly. It is as if this movement switches on the sound.

The cry of a curlew rising from the edge of Llyn Edno on the way back reminded me how silent the September hills are compared with April when curlews are bubbling and wailing and there is endless lark-song overhead. Now the only sound as I returned in the fading light was the distant faint notes of a flock of pipits among the rocks above Llyn Llagi. From the half-dried-up pool called Biswail a green sandpiper went silently away, probably one of the birds I had disturbed at the quarry reservoir that morning, for this is a scarce bird of passage in the hills. I crossed the ridge of Cnicht half a mile north-east of the summit. Cnicht is deservedly one of the more popular lesser mountains of Snowdonia, its sharp ridge being just the height for a day's scramble with the family.

As a view-point it is excellent; and as part of the view from the Portmadoc side it looks superb for then it is end-on and appears as a slender peak, 'the Matterhorn of Wales', as the guide-books have it. In fact Snowdon has a better claim than Cnicht to be called the Welsh Matterhorn, for Snowdon is a peak from nearly all angles and, like the Matterhorn, is the product of cirques that have cut into the mountain from all sides to produce a sharp summit. As I left Cnicht and came round the head of Cwm Croesor six choughs, flying in pairs, passed over towards the east calling beautifully. In a few minutes there came ten more, also in pairs, on exactly the same line. I watched them go out of sight into Cwmorthin to some communal roost down there or in the cliffs nearer Blaenau Ffestiniog.

I slept that night on the slopes of Moelwyn in the hope of seeing a return flight of choughs at dawn. Before midnight all haze had gone from the sky, the stars became brilliant and the body of the Plough, the right way up at this season, hung poised close over the top of Cnicht. Next morning before the last stars had faded, a ring ouzel, the only one I saw that week, chattered at me from a few yards' range, unsure what I was, lying there shadowed by a rock. Three ravens passed in silence except for the *whoo-whoo-whoo* of their wings. But there was no return flight of the choughs.

I made my way down to the cross-lanes at Croesor, turned right through the village and followed the track round to the Nantmor valley. By mid-morning the sun was warm. Lizards were active along the walls. Robins sang their autumn song from the thorns. An almost black adder uncurled itself in the middle of the road and slid quickly into the heather. It was a perfect September morning spoilt momentarily by a screaming, low-flying jet-bomber which shadowed my mind with sad and ugly images. But the sight of the Nantmor valley dispelled all gloomy thoughts; lovely Nantmor with its woods and arid rocks; its secret hollows and side-valleys; its old houses and farms; its clear, cold, green-pooled stream; its essential wildness. I remembered the year I once lived in this valley: the long dry sunny spring full of the songs of pied flycatchers, redstarts and grey wagtails; the unending rains of August and half September; the quieter days of autumn. I thought of the eggs in the ring ouzel's nest found in the heather going up to Llyn Edno in April; of the tiny spiky caterpillar I picked up one day in mid-May in the gorge below Cae Ddafydd and which, reared on violet leaves in a jam-jar, eventually became the loveliest butterfly I ever saw, for there cannot be anything more delicately beautiful than a freshly emerged silver-washed fritillary. And I remembered October mornings when flock after flock of fieldfares, redwings, bramblings and chaffinches passed down the valley to the south; and how I traced their flight-line backwards and saw them also at Ogwen flying east through the pass between Pen yr Oleu Wen and Tryfan.

I spent the rest of the day pottering among the many rocks and bogs of Arddu. Surely only the flanks of the Rhinog above Cwm Bychan are barer than the terraced ridges of this side of Nantmor? At Llyn Arddu were more of the big golden-ringed *Cordulegaster* dragon-flies, now very close to the end of their season and their lives. By a stream on my way down, the twigs of a low-growing willow were so bereft of leaves only a hawk-moth caterpillar could have eaten them. Nor did I have to search long before finding a fine, full-grown eyed hawk, pale-green, white-striped, blue-horned, eating into a leaf at phenomenal speed. This seems the commonest hawk-moth of these higher willows though perhaps the poplar-hawk is as numerous. Two or three swallow-prominent caterpillars, conspicuously bright green on the grey leaves, and also the strangely humped pebble-prominent caterpillars, were feeding on the eared willows down this bouldery stream-side. That evening I made my way round to Capel Curig to make a start on Moel Siabod next day.

I know of no easier approach to Moel Siabod than the one from Pont Cyfyng up the slate-quarry road. In the cool oakwood a chiffchaff sang and I wondered if it was the last of the year. By the farm above the wood swallows were in rows along the wires, quietly preening (do they ever stop preening?) and very tame. Much yarrow was in flower about this farm, betokening some kindness in the soil. Above the farm two miles of heather-moor between there and Siabod looked formidable, the sun being already hot. Yet nothing could have been more gentle, for the quarry road continues straight to the foot of Siabod, a long-abandoned road, now a green road with a short dry turf delightful to the feet. On my right through the inevitable haze rose the grey peaks of the Carneddau, and below them the wide hollow of rocks and bogs up which the track goes from Capel Curig over to Crafnant. Thankfully I found a cooling, scented wind coming across the heather from the south. An autumn St. Mark's fly settled on my sleeve, a long-bodied, blundering, harmless creature with handsome black legs that were reddish-brown on the joint nearest the body. A skylark chirruped past, the first I had seen that week, for though countless larks nest on the hills they nearly all depart as soon as breeding is over.

The heath rush, abundant along the edges of this green road, was heavily infested with the rush moth (*Coleophora caespititiella*), whose caterpillars feed on its seeds. Every plant I examined had at least two or three of the whitish larval cases, about a quarter of an inch long, that stick out so conspicuously on the nut-brown seed-heads of the rush. What is always worth observing is the height at which this infestation ceases, for though the rush continues up the mountains practically to their tops, the moth is far less of a mountaineer and you usually find that its larval cases die out at somewhere about 1,500 feet, though they have been recorded up to 2,000. Along my road that day they ceased abruptly at the quarry reservoir (about 1,300 feet). Of two other rushes also growing along the track, the common rush was very little affected by the moth and the sharp-flowered rush not at all.

The bracken on the slope above the reservoir had turned a pale yellow and there was a richer yellow down the rocks where the western gorse was in bloom. This gorse, *Ulex gallii*, is far hardier than the spring-flowering lowlands gorse, *Ulex europaeus*, that gets so cut back in severe winters. Here on these rocks the western gorse grows square to the east wind and revels in it. Near it an aspen had sprouted out of a crack in a hard face of rock. Not, you would think, the most promising of sites for any tree. But the aspen likes mountain rocks. You can see a little of it on the ledges of Cwm Idwal where it rustles almost incessantly in the breeze. Perhaps, like the wood anemone, the aspen does not like the sun, for the anemone also grows in shady woods and on cool mountain ledges. This Siabod aspen had been there some years. Its trunk was thick and sturdy but its height was no more than five feet, probably the greatest height possible for an aspen exposed to a maximum of wind with its roots in a minimum of soil. I climbed up to look for caterpillars of poplar-hawk or puss; but not a leaf of it had been nibbled. For trees there are compensations in living the hard life of the uplands.

The track now wound up through slate-tips and derelict buildings. As for the tips, I can accept them provided they do not dominate the landscape as they do at Blaenau Ffestiniog. A few small slate-tips are only as if the mountain had vomited an excess of scree; and the woolly-haired moss and parsley fern do their best to make them look natural. Hereabouts the rocks are most delicately coloured by lichens: green, white, red-brown, grey and black. Some of the patches of colour are edged round with wavy black lines and all the patches meet in strange and beautiful patterns. A bright yellow-green lichen alongside a red-brown one was particularly striking. These rock surfaces, though the lichens looked so fixed and static, are really lichen battlefields where one species relentlessly ousts its neighbours and is itself ultimately elbowed out of existence: a very slow process for lichens are almost everlasting.

Above the reservoir you enter a cwm and soon the track goes up a green incline between slate-tips and ends suddenly at a sheer edge with water yards below. This is the old quarry, now flooded—a frightening, deep, square pool with sheer, smooth rocks all round. Its north-facing wall is worth examining: it is a huge vertical face on which the structure of the slate is clearly exposed. Siabod, then, is a mountain fundamentally of slate, of sedimentary

Ordovician rocks like all the neighbouring heights. But because of its hard igneous cap it has endured while deep valleys like those of the Lledr and the Llugwy have been carved around it. Now, at this quarry, you see, if you have not realised it long since, that even Siabod, which seems to confront the world on all sides with such stolid compactness, has its point of weakness and is deeply penetrated by this cwm that cuts right into its heart from the north-east. It is in fact a hollow mountain with an arc of cliffs inside it and a lake below: the pattern of so many other mountains.

At the quarry all roads cease and you are left to find your own way through a mile of a grassy pass between rocks to the edge of the lake. Llyn y Foel is a lake of rocks. All its banks are naked rock; there are small rocks sticking up in mid-water; and there are two larger islands, one of them far enough off-shore to be the sort of place black-headed gulls enjoy nesting on. West of the lake rise the great walls of Siabod. The eye travels up them to as rough and broken a skyline as on any mountain in Snowdonia. No one needs to be told that Siabod, though not now reaching quite 3,000 feet, has towered much higher. For vast ruins of rocks lie crashed in the bottom of this cwm, countless jagged blocks of rough-surfaced dolerite, graded in size from the smaller ones halfway down the slopes to the huge block scree at the bottom. Detached from the main cliffs, but not yet fallen to the abyss, stand broken-off stacks and pinnacles that are obviously mere worn-down stumps of something once gigantic.

It was over these rocks, with three unwelcoming ravens croaking vibrantly above me, that I made my way up the scree, following the stream, as botanists inevitably do. Here under the crags was stillness. There was no noise of wind. The humming of bumble-bees in the heather reached me from yards away. And now the mountain plants began to appear. Green rosettes of starry saxifrage leaves, then a few plants still in flower. Then crowds of tiny white flowers on extremely long brown stalks—the New Zealand willow-herb. I suppose this invader is still spreading and is destined to decorate every streamside of upland

Snowdonia. Its abundance up this scree and on Moelwyn is impressive: yet I do not recall it on Hebog. The pretty yellow sedge, its fruits now full and ripe, hung stiffly over the water. Higher up were abundant mats of leaves of the golden saxifrage, a plant already with its thoughts on spring, for few species flower earlier in the year. There were several other of the usual species that grow on wet rocks: common lady's mantle, lesser clubmoss, many butterworts and mossy saxifrage. And on dry rocks the English stonecrop still flowered well. What an admirable plant this is that spreads its white stars on the thinnest, most miserable soils from the sea-cliffs up to nearly 3,500 feet, as on Carnedd Llywelyn.

The cliffs of Siabod are mostly poor in plants. It seems to be only the stream down this particular scree that has discovered any calciferous rocks in all that vast cirque of crags. These richer rocks are unmistakable when you reach the top of the scree, for there they are directly above you, bristling with the fleshy stems of rose-root which in mid-September were turning colour and giving a pink look to the cliffs. Ledgefuls of ox-eye daisies were in full and splendid flower. Extra-large harebells nodded gracefully. The showy purple-blue heads of devilsbit scabious hung out into space. But now I turned away from these wet, brown rocks and took to a dry, scree-filled gully to get to the top; and in so doing I missed the purple saxifrage which I ought to have found there, as Evan Roberts told me later. Instead I scrambled several hundred feet to the top, seeing no more plants except a lone juniper.

The inevitable meadow pipits shared the summit rocks with me for it was quite warm up there in the sunshine. How variable these pipits are. Some were dark-backed, heavily pencilled on the breast and quite lacking a white eye-stripe. Others were almost pale brown, and only lightly streaked on the breast and with pronounced pale eye-stripes. Delightful, lively birds that I never tire of seeing, though so common.

Siabod, like all detached mountains, is a good place from which to see the rest of the Park and a lot more besides: but you need the clarity that brings or follows rain. Then you have not only vast views all round the east and south but, very close on the west, you can look straight into the Snowdon Horseshoe with the morning sun at your back. An even more intimate view of the Horseshoe, especially useful for photographers, can be got by following the western ridge of Siabod two miles to the hills above Llyn Diwaunedd. But during an intense anticyclone you will see little even from Siabod. All I could make out that September day was a shadowy outline of Moelwyn and a few mysterious glimpses into the Lledr valley through occasional rifts in the murk that revealed plantations of conifers whose straight top-edges quarrel so violently with the natural roll and curve of mountain landscape. The strung-out houses of Dolwyddelan shone palely to mark the line of the main road onwards towards the Crimea Pass. But it is the old roads of that side of Siabod that are best to follow if you are a walker. There is the reputedly ancient path from Llyn Gywnant to Dolwyddelan Castle; and there is Sarn Helen which picks its way shyly over the hills from Betws-y-coed and up the fine valley of Cwm Penamnen, then over the top and down into Merioneth east of Manod. In at least some sections this part of Sarn Helen is a well-paved Roman road. All that country where the slate region merges into the wet moors of Migneint is a fascinating area whose natural history is little known. It is a rough, bare land full of old quarries (and choughs) and reservoirs, broken slopes, deep little valleys, and sudden rocky hills with sharp crags, some lime-rich and with calcicole plants such as green spleenwort in the cracks. A small but dramatic area.

As I came down off Siabod that day following the long valley of the quarries, two brown wheatears flicked away in front of me, the only ones I had seen anywhere on Hebog, Moelwyn or Siabod. What clearer sign was there that winter would soon be creeping up into the hills than the sight of the year's last wheatears? And now in the evening the sky had turned from grey-blue to a deep and clear blue, and wild waving feathers of cirrus cloud were quickly forming all over it. The wind was rising and sending long white streaks down the length of the quarry reservoir. By tomorrow the fine weather would be ended.

In these few days I could not go everywhere nor see everything, and there are many folded valleys, woods, lakes and crags I have not mentioned. There is Llyn Elsi in the east, a small moorland lake whose character is changing, alas, as the conifers grow up at its edge; a lake of great charm with its rocky walls, its black-headed gulls, its floating plants, its tree-clad bluff, its islands, its sundew-rich bogs, its mountain views. In complete contrast to Elsi there is Llyn Cwellyn in the west, reflecting its massive bare hills, a fine lake over a mile long and very full and deep. There are the miles of high crags along the south slopes of the Nantlle valley and many more on Moelwyn Bach and Moel yr Hydd: all inviting places that can be explored and re-explored by naturalists without exhausting the possibility of new discoveries, new delights. And there are the rivers. In the east the Llugwy, the Lledr, the Machno, the Conway and all their side-streams pouring off the moors down their slides and waterfalls and uniting in the wooded gorges that are the charm of the Betws-y-coed area. In the west there is the Glaslyn: and the shades of all previous travellers would haunt me for ever if I said no word of Aberglaslyn Pass. It is, of course, a splendid place with its river, its bridge, its spectacular cliffs, its high-perched conifers. It is one of the few places that are best visited during long and heavy rains when the river comes down the gorge in a roaring brown and white flood. Yes, winter is best for Aberglaslyn Pass. In summer it can be just another place with a traffic problem.

Finally, there is the lower Glaslyn valley where the river is wide and slow and tree-fringed; and mixed woods, some of the finest-looking woods in Wales, climb up the lower slopes of Hebog, woods that are particularly good for an autumn foray to see the fly agarics, chanterelles, honey tufts, death caps, beech tufts, stinkhorns, parasols, puff-balls, boleti and the many other fascinating species that make up the rich world of the woodland fungi. The river leaves these woods to follow the course of its former estuary to the sea. You can if you like side with Shelley and praise William Maddocks for turning this estuary into land. Yet surely estuaries such as Mawddach and Dovey are among the chief glories of Cardigan Bay? They too presumably could be turned into dry land, and any schemes to do so could be defended on utilitarian grounds. But there must be some limit at which the most rabid philistine will pause. Someone has pointed out that no one would dare suggest demolishing a beautiful cathedral to make way for a block of flats. Why then destroy a beautiful estuary? I like to think that some day a generation will arise with the imagination to replace Maddock's embankment by a fine bridge and let the tides reclaim at least part of the old Glaslyn estuary.

CHAPTER NINE

From Arennig to Rhinog

The two peaks of Arennig, with three miles between them, stand up splendidly from the moorlands all round, especially, I think, when you see them from some high point of the north end of the Rhinog range such as Moel Ysgyfarnogod—the Hill of the Hares. There are many differences between the two Arennig heights. The southern one, Arennig Fawr, rising to 2,500 feet, is not only much higher than Arennig Fach and bulks far bigger in the landscape, it is also more craggy and rock-strewn and has more mountain personality about it. But they have points in common: they both tower up because their hard igneous rocks have endured while thousands of feet of once enveloping softer rocks have been eroded from about them; and they both on their north-east flanks break off into high cliffs that hang over clear and beautiful lakes. The Arennig country, composed as much of it is of both rhyolitic and doleritic rocks, has the inevitable mixture of acid and basic soils. But the effect of these differences on the vegetation appears to be small, for calcicole plants are not much in evidence. Not that it is a region that has been intensively botanised and some good finds may be made there yet. It is possible that the melancholy thistle grows somewhere on Arennig for some years ago a party of us found one plant of it by Arennig station (1,140 ft.). A railway plant could have come from anywhere but perhaps it grows in a natural habitat not far away. This thistle is a very rare plant in Snowdonia, which is perhaps a pity for it can be used, it is said, as a cure for melancholy and so might be very useful to botanists after a disappointing day in the field! It was recorded on

Snowdon itself by the earliest botanists but is now unknown there.

Llyn Arennig Fach is said to swarm with small perch, which is not a common fish to find in an upland Welsh lake. But what impressed me about these lakes were the numerous shoals of minnows in the shallows: so I was interested to see that Forrest was also struck by them at the beginning of the century: 'The largest and most gorgeously coloured minnows I ever saw were in Llyn Arennig [he does not say which lake]. This was in May, 1902, at which time they were spawning, and so crowded together in the rivulets running into the lake that I scooped numbers out with my hands. Many of the females measured over four inches long. In June, 1900, I saw countless thousands at the edge of Talyllyn Lake.'

To go to Arennig Fawr in May, as Forrest did, is to go at a delicious time of year. The approach from the south from the village of Parc is particularly delightful just then when the slopes are loud with cuckoos, and countless pipits are coming down the sky in song; by the last habitation, the lovely old stone cottage of Cefn-y-maes, the beeches are a fresh new green, and here and there if the sun is out the little green hairstreak butterflies rise from the turf in front of you, their underwings flashing brilliantly emerald. A fine prospect rises on the south as you get higher, an undulating mountain line from Berwyn and Aran all the way round to Cader Idris in the west. It is all wonderfully wild country as long as you do not look north-east and see the unpleasant line of pylons that crosses the hills conspicuously there. You squelch over a wide, boggy, curlew-calling valley with a peat-red stream flowing through it where whinchats sing and display among the rushes. Then more scrapey songs as you go up through the rocks beyond, and this time it is the wheatears that are chasing excitedly about you. You soon come to the cliffs that stand in a high, broken semi-circle above the lake. It is a corrie of deep silence and peace, a place to listen all day to the piping ring ouzels and to expect a pine-marten to appear among the moraines. Here the ledges are covered with the vegetation of acid soils. There is much heather, crowberry,

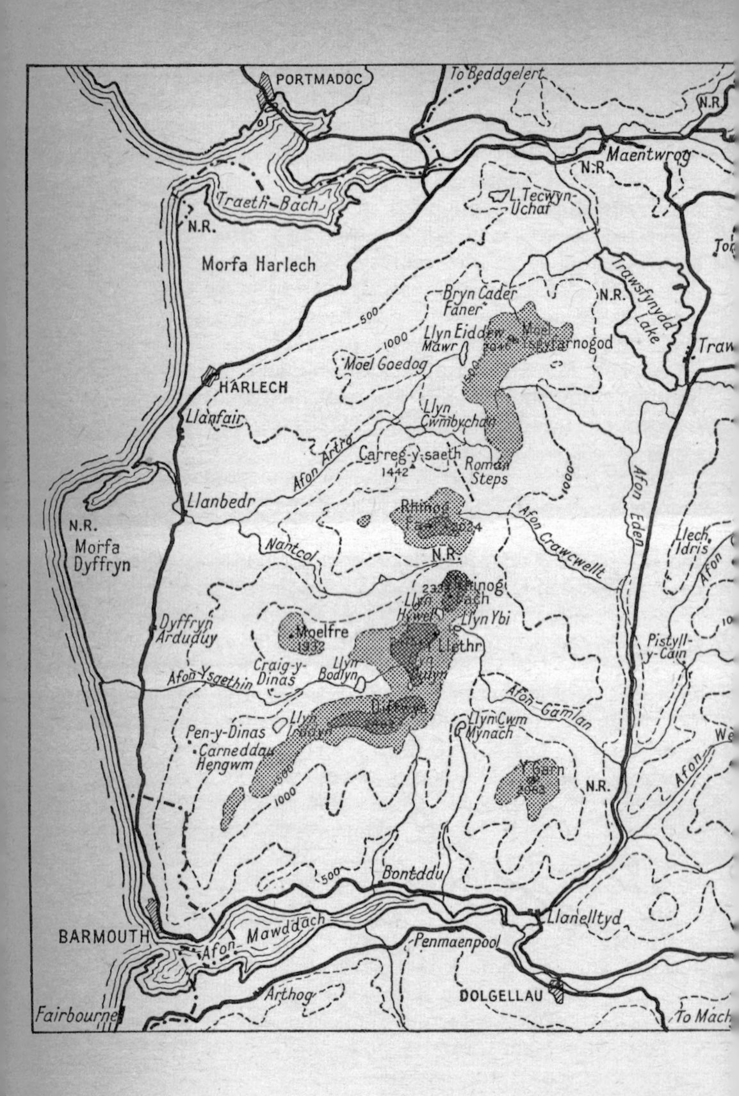

PORTMADOC
To Beddgelert
N.R.
Maentwrog
N.R.
L. Tecwyn Uchaf
Traeth Bach
N.R.
Morfa Harlech
Trawsfynydd Lake
Bryn Cader Faner
N.R.
Llyn Eidden Mawr
Moel Ysgyfarnogod
Moel Goedog
HARLECH
Llyn Cwmbychan
Llanfair
Afon Artro
Carreg-y-saeth
1442
Roman Steps
Afon Eden
Llanbedr
N.R.
Morfa Dyffryn
Nantcol
Afon Crawcwellt
Llech Idris
N.R.
Rhinog Fach
Llyn Hywel
Llyn Ybi
Dyffryn Ardudwy
Moelfre
Pistyll-y-Cain
Craig-y-Dinas
Llyn Bodlyn
Afon Ysgethin
Afon Gamlan
Pen-y-Dinas
Llyn Cwm Mynach
Carneddau Hengwm
Y Garn
N.R.
500
1000
1500
Bontddu
Llanelltyd
BARMOUTH
Afon Mawddach
Penmaenpool
Arthog
DOLGELLAU
Fairbourne

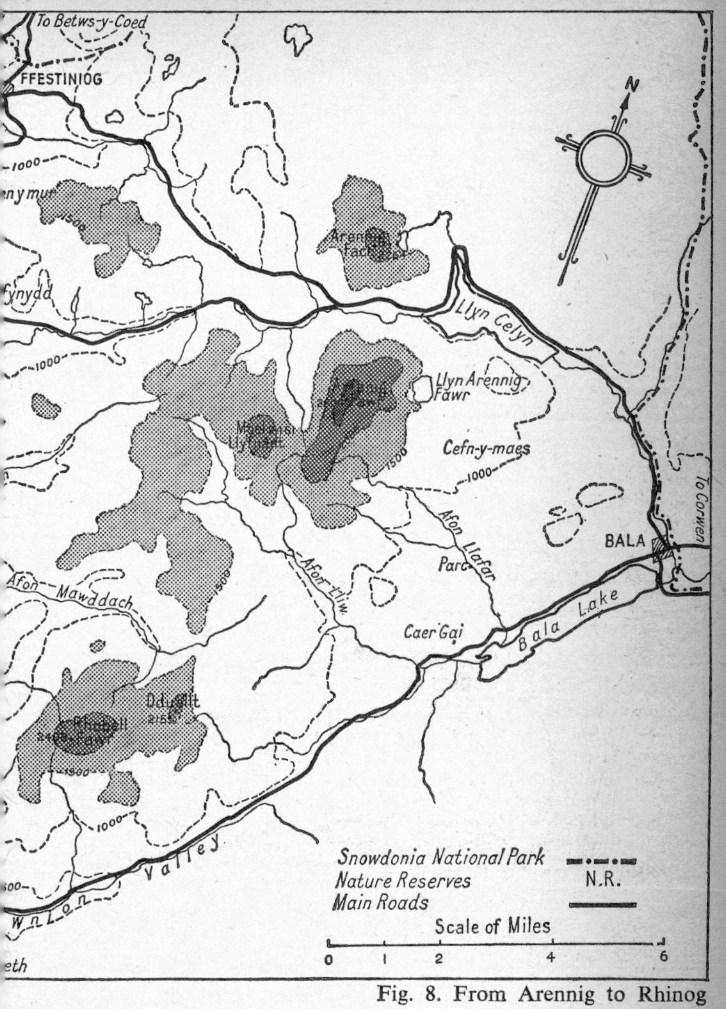

Fig. 8. From Arennig to Rhinog

cowberry and filmy fern; and bumble-bees roar endlessly among the bilberry flowers. Beech fern has beautifully arrayed itself in some angles of the rocks (does any fern do it better?) and the largest wood anemones I have ever seen were on one high ledge half a mile south-west of the lake at nearly 2,000 feet. But it is quite characteristic of this anemone to flourish on such high, wet rocks where it never sees the sun. These cliffs of Arennig were among the places where wrens survived the arctic winter of 1962–3, though they were wiped out over large areas of the lowlands. For the first time on record one or two pairs of herring gulls began to nest around these lakes in the early 1960's.

Between Arennig Fawr and Arennig Fach flows the Tryweryn stream, dammed in the early 1960's to form a reservoir for Liverpool, a work that destroyed the hamlet of Capel Celyn. So, alas, a fragment of old Wales disappeared. But reservoirs can bring real beauty to a place and can be a refuge for wild duck and other birds that are these days being increasingly persecuted by sportsmen. Along this valley there used to run the railway from Ffestiniog to Bala, a remarkable high-level line that wound across the moors for miles east of Trawsfynydd. I have vivid memories of going by train up the Prysor valley one winter day when all the mountains and moorlands were brilliantly white in the February sun and Llyn Tryweryn was a gleaming sheet of ice.

Two other Arennig streams, the Llafar and the Lliw, have so far escaped exploitation. The Lliw joins the Dee just above Bala Lake and close to the Roman fort of Caer Gai; the Llafar flows directly into the lake. The Lliw valley takes you seven miles north-west up to the wet moorland that lies beyond the sharpish symmetrical hill called Llyfnant, which, like Arennig, has some calcareous rock and may possibly yield some interesting plants not yet recorded. As an encouragement I could mention that on rocks in this area the serrated wintergreen (*Orthilia secunda*) was discovered in 1961, its only known Snowdonian locality. Outside Scotland, this is an extremely rare plant in the British Isles. But unless it is flowering it might easily be missed, for it is rather small and its leaves could be mistaken for those of common speedwell, which also grows on mountain ledges.

If you have a taste for real moorland and do not mind a weary slog across tussocks of cotton-grass you should turn off the Lliw valley at Castell Carndochan and make your way south-west up the narrow heathery vale of Nant-y-fign. But go up past the old gold-mine to look at the *castell* first. From a distance it looks like an Iron Age hill-fort, but when you get to the top of its bleak knoll you see it is in very truth a castle with quite substantial foundation walls of squared stone in whose mortar that enterprising fern, the common spleenwort, has planted itself. The mortar is of interest. Lhuyd, who came here in the early 1690's, gathering material for *Britannia*, says the lime for the mortar was made from cockle-shells 'which must have been brought hither by land carriage about 14 miles' from the Mawddach estuary. To find the words *castell* (castle) and *carn* (cairn) in one name is unusual but the ruinous scattering of small stones also on the summit of this rock suggests that a large cairn stood here from prehistoric time. History has little to say about this castle and Fenton, who saw it in 1798, was probably right in concluding it was 'the refuge of some ferocious chief who had lost all claim to society and who lived by plunder and rapine'.

You are barely two miles up the narrow valley of Nant-y-fign before you find it widening out into a quaking morass with the remains of a now moss-choked lake in the centre; which is proper enough since *fign*, or *figyn*, is a form of *migyn*—the Welsh word for a quagmire. Any place marked *figyn* on the map is a likely breeding place of teal, for they love the thickly concealing rushes you always find in such places. But you can waste a lot of time trying to raise teal from such a stronghold, especially when they are reluctant or unable to fly during the late-summer moult. Besides, there is further allurement ahead, for from the banks of the Fign you see the high rocks of Dduallt, the Black Slope, begin to peep over the moor. Dduallt is a short, east-facing ridge with little in the way of precipices though it has some interesting wet gullies. There is by no means a rich flora but that there is some lime in the rocks is attested by the presence of plenty of mossy saxifrage, a little burnet saxifrage and bladder fern, and more oak fern than I have seen anywhere. Oak fern here grows on the rocks, it springs out of the screes and—something I never saw anywhere else—it even grows through clumps of bog moss among the heather. A plant that is curiously rare on these rocks is the starry saxifrage. But an animal that is certainly not rare is the field vole. I have never seen so many vole runs on an open mountainside as were to be seen on the ledges and slopes of Dduallt in September 1963. Clearly it is not only when ground is enclosed for forestry that vole numbers multiply exceedingly. Not surprisingly buzzards, kestrels, ravens and a fox were on the rocks of Dduallt that day.

Many mountain tops are claimed to have the best view in Wales. But if you want a central height from which to see the other mountains of Snowdonia you can scarcely choose better than Dduallt or its taller neighbour Rhobell Fawr. From Dduallt you have the Caernarvonshire and Merioneth mountains all round you in one uninterrupted circle. Immediately below you on the east a widely stretching cotton-grass moor rises to a slightly domed watershed marked by deeply eroded peat hags whose channels are scattered with the remains of prehistoric birch trees. Three streams flow south from this watershed, three rills through the heather that quickly unite and become the River Dee which wanders a mile or so through quaggy ground in a hesitating, southward way before, with sudden resolve, it turns east for Bala Lake.

Although Dduallt looks imposing as you near it from the east, yet when you get to the top and look round, you see that it is really Rhobell Fawr that is the dominant mountain of this part of the Park, for its slopes and foothills reach across many square miles of largely uninhabited country and send off many streams to feed the Mawddach or the Wnion. Geologically Rhobell Fawr belongs with Cader, Aran and Arennig to the Ordovician igneous summits: but its structure is complex even for

Snowdonia and I once heard a geologist say its rocks are as confused as if they had been tipped there out of a wheelbarrow. Not surprisingly, there is a complete mix-up of acid and basic rocks. But as on Arennig and Llyfnant and so many other mountains, the rocks over wide areas are so deeply masked by peat that the vegetation rarely feels the benefit of what mineral wealth may lie in the underlying rock. Only on the south-east, where the little torrent called Eiddon has cut a gorge and deeply laid bare the sides of the mountain, does the truly calcareous nature of the rock come to the surface. Here at Craig y Benglog is a surprising gathering of calcicole plants to find among the acid hills of Snowdonia. When I say surprising I do not mean rare, as you will see when I say that one of the astonishing plants of Craig y Benglog is none other than the common rock-rose, a plant so ordinary in chalk or limestone districts that no one gives it a second look. But to see its showy yellow flowers on the rocks of Snowdonia is truly rare, for it grows nowhere else in the Park except very locally round the skirts of Rhobell Fawr and between Trawsfynydd and Arennig.

The ecology of Craig y Benglog was the subject of excellent work by Price Evans (see page 91). With outstanding success Price Evans followed up the work of the geologists A. H. Cox and A. K. Wells by showing how intimately the plant-rich localities are associated with the basic volcanic rocks that had been discovered and described at Craig y Benglog. The most striking botanical feature of Benglog is the ashwood that clothes the scree below the rocks, for a mountainside ashwood in the Welsh uplands is rare and a sure indicator of base-rich soil. Besides the ash trees and the rock-rose there are the usual two calcicole mosses, *Neckera crispa* and *Ctenidium molluscum*, as well as:

Hairy rock-cress	*Arabis hirsuta*
Fairy flax	*Linum catharticum*
Early purple orchid	*Orchis mascula*
Burnet saxifrage	*Pimpinella saxifraga*
Shining cranesbill	*Geranium lucidum*

Marjoram	*Origanum vulgare*
Spindle	*Euonymus europaeus*
(here obviously native as distinct from the spindles near old cottage gardens planted by the peasant spinners)	
Small-leaved lime	*Tilia cordata*
Wych-elm	*Ulmus glabra*
Rock stonecrop	*Sedum forsteranum*
Wood spurge	*Euphorbia amygdaloides*

The wood spurge is a surprising plant to find up here in a mountain gorge for it is normally a species of lowland woods. It grows on similar base-rich ledges in the upper Dovey valley but is not recorded elsewhere in Snowdonia. Price Evans complained that heavy grazing in the Benglog ashwood not only by sheep but also by twenty or thirty goats was preventing regeneration of the trees. These 'wild' goats are still there, grazing down among the trees in winter and ranging far over the mountains and much harder to find in summer. They are among several herds of these fine-looking but destructive animals that range the Snowdonian uplands. On dry ground above the Benglog ashwood I have found a little of another calcicole, the mountain everlasting, but this species, in Snowdonia close to its southern limits, is rare here and usually in very small quantity. There is a little along the Cader range and at one place on the Merioneth coast. Lhuyd reported it as abundant in 1682 near Bird Rock (Craig yr Aderyn) in the Dysynni valley: 'On a mountain called Cefn Llwyd, ye back side of Craig yr aderyn . . . in flower, plentifully.' This locality is now lost but perhaps the plant will be found there yet.

The Mawddach is the longest and best-known river wholly within the National Park. Many Snowdonian rivers spring out of mountain crags, as does the Dovey from the rocks of Aran Fawddwy; but the Mawddach, like the Dee, rises in the cotton-grass moors east of Dduallt. While the Dee slips away south-east, the Mawddach creeps northwards out of the other side of the same dome of peat, soon to gather many other springs from all over the spongy tract called Waun-y-griafolen—the Moor of the Rowan

PLATE XVII *Above*, Llyn Tegid, or Bala Lake, looking south-west to Aran Benllyn (2,901 feet). The lake lies in Snowdonia's biggest fault-line, and is noted for its fish, especially the gwyniad which is found nowhere else in the British Isles. *Below*, Llyn Mwyngil, or Talyllyn Lake, lies in the same fault-line under the southern slopes of Cader Idris

PLATE XVIII *Above left*, the wheatear — a male is pictured here — is a characteristic bird of open rocky places, nesting in walls and in holes under boulders. *Above right*, like the wheatear the common sandpiper is a widespread summer visitor that nests from near sea-level up to high altitudes. It is commonest by moorland streams and upland lakes

Right, the chough is an uncommon resident, breeding from the coast to the mountains but is most frequent in the vicinity of Merioneth slate quarries

PLATE XIX **Sea-Birds Inland.** *Left* at Bird Rock near Towyn, Merioneth, cormorants breed on a cliff four miles from the coast. *Below*, herring gulls at 3000 feet along the Snowdon Horseshoe Track. In recent years this gull has learned to hang about picnicking parties for scraps of food

PLATE XX **Birds of Young Conifer Plantations.** *Above*, newly planted forests are often infested with voles which attract short-eared owls. The owls nest on the ground, often under the cover of the little trees *Below*, the lesser redpoll is a very successful coloniser of young forests and is particularly fond of nesting in Sitka spruce, as here

PLATE XXI **Woodland Birds.** *Above left*, willow tit, a thinly scattered resident species of the edges of boggy thickets. *Right*, a pair of pied flycatchers. This is a common summer visitor breeding in holes in trees in deciduous woodland
Below, the sparrow hawk found throughout the woodlands of Snowdonia

PLATE XXII *Above*, gwyniad from Bala Lake. A small member of the salmon family, the gwyniad is closely related to the several other species of British whitefish (*Coregonus*) but has become distinct from them by long isolation in Bala Lake. Feeding on minute forms of life it is seldom caught on rod and line. Another of the Salmonidae, the char (*below*), though less restricted than the gwyniad, is found in only a few of the deeper Snowdonian lakes

PLATE XXIII *Above*, the north face of Cader Idris, a fruitful area for ecological studies because of the variety of its rocks. The top of the escarpment consists of a hard sill of granophyre, an igneous rock
Below, the estuary of the Mawddach at low tide. The mountains on the horizon are Arennig (extreme left), Rhobell Fawr (centre), with Aran Benllyn and Aran Fawddwy on the extreme right. The low-lying ground (bottom right) is Arthog Bog with the foot-hills of Cader Idris beyond

PLATE XXIV Sunset over Llyn Cwellyn and the slopes of Mynydd Mawr. Cwellyn is deep and is one of the few Snowdonian lakes which contain char

Tree, a name which, like so many others, recalls trees that no longer exist. From this grouse-barking moor that has in it another *figyn*, the quaky-edged, decreasing pool called Crych-y-waen, the Mawddach drops suddenly over the rim into Cwm yr Allt-llwyd, a steep-sided valley where, if you have plodded across the moors from Pennant Lliw, you find the first trees and the first houses you have seen for several hours.

Trees, but not houses, increase quickly down Cwm yr Allt-llwyd and in a few miles the river flows into deciduous woodland. But this pleasant oakwood is, alas, a mere fragment of similar forest that once filled all the valleys here but is now almost entirely replaced by the conifers of Coed-y-brenin. But the waterfalls are still there, the three famous falls that were compulsory visiting for the tourists of years ago: the falls of the Mawddach itself and, very close by, the more splendid Pistyll y Cain; and then a few miles downriver the Rhaeadr Du on the side-stream called Gamlan that comes down off Rhinog. Where the Cain flows into the Mawddach is a place of much delight, a place of roaring waters shadowed by oaks and rowans, a place of rocks, moss, ferns and slender St. John's wort; of buzzards, grey wagtails, pied flycatchers and large wood ants scurrying about the gold-mine ruins. To see it at its most spectacular, go after a day of torrential rain and learn at first hand about the cutting, landscaping power of rivers. In 1798 the Reverend John Evans, in pouring out the inevitable raptures about the falls, describes the valley sides as 'dark with the umbrageous shade of beech and oak'. Though we may wonder how shade could be other than umbrageous, it is interesting to find these valleys covered with mature beech at that time, for beech is not a native tree in North Wales and was evidently established there during the great enclosures and plantations of the eighteenth century.

The history of these extensive Mawddach woodlands is bound up with the history of two ancient local houses, Hengwrt and Nannau. At Hengwrt in the seventeenth century lived one of the greatest of antiquaries, Robert Vaughan, who collected and copied a vast number of scarce and valuable old Welsh manuscripts (including some of Edward Lhuyd's), a priceless collection now kept for posterity by the National Library of Wales. It was a descendant of his, Sir Robert Vaughan, who laid out the Nannau estate as a rustic paradise, as Thomas Jones had recently done at Hafod in Cardiganshire, planting prodigious numbers of trees. But it is the native oak that is the naturally dominant species of all these valleys, and some famous ones are on record from Nannau and Ganllwyd. One of 609 cubic feet was felled at Ganllwyd about 1790 and was 'so remarkably large', says Cathrall, 'that it was denominated Brenhinbren, or the King Oak; and it was celebrated in the songs of the Welsh bards'.

The only monastery in Merioneth at the time of the Reformation was the Cistercian abbey at Cymer near where the Mawddach and Wnion rivers meet. It was a monastery that wielded great influence over a vast area for several centuries after 1209, when its charter gave it possession of estates stretching from Barmouth inland to beyond Dolgellau and including 'rivers, lakes and seas; birds and wild and tame beasts; mountains and woods; things moveable and unmoveable under and over the lands so granted; also the liberty of digging for metals and hidden treasures'. It can be assumed that such talented and determined exploiters of natural resources as the Cistercians would take full advantage of the privileges they had been given. We know that their brethren at Strata Florida abbey in Cardiganshire mined deeply for lead and pastured great flocks on the mountains, and doubtless monastic life at Cymer ran on similar lines. Many of the innumerable trial borings still visible in the rocks of the mineral-rich Mawddach region were presumably made by the monks of Cymer.

Just below Cymer the river meets the salt. Here at Llanelltyd the monks would have netted the salmon, built boats and commanded an access to the sea, for Llanelltyd, like Penmaenpool two miles downstream, has been an important place on the river. It was from Llanelltyd that Lhuyd received the only note his questionnaire produced for him on the subject of choughs. It reported that someone in the village possessed a tame but unknown bird that was like a jackdaw but had red legs: an interesting indication that the chough was probably about as rare in the seventeenth century as it is now, for it was evidently just as unfamiliar to people.

The Mawddach estuary does not need me to sing its praises for it is justly celebrated as among the loveliest spots on earth. But it was not so well known to the earliest tourists because through most of the eighteenth century Barmouth was reachable from Dolgellau only by a hair-raising mountainside road. So Barmouth, such as it was before sea-bathing days, was usually by-passed, the travellers turning north-west at Bont Ddu along the ancient road to Harlech. This is only one of countless fascinating old trackways over the hills of Snowdonia. Many are now hard to trace where they have sunk into soft ground or are overgrown with rushes or bracken. But all are worth trying to follow for the interesting country they lead you through. Readers of Welsh have the advantage of being able to consult several excellent accounts written in Welsh about various towns and parishes in the Park, for the Welsh people are assiduous producers of local histories. In Merioneth for instance there is *Hanes Dolgellau* (A History of Dolgellau) by John Jones, 1872, and *Hanes Plwyf Llanegryn* (A History of Llanegryn Parish) by William Davies, 1948. And there are many other excellent parish and town histories available.

Though hemmed in by mountains on either side, the Mawddach estuary has room for an extensive stretch of bog near Arthog. This long ago was no doubt a perfect raised bog, in the botanical sense, being slightly domed in the centre. But virgin bogs are extremely rare in Wales now because of their long exploitation for peat, and Arthog bog, so close to the estuary, was particularly vulnerable, its peat being sought by boat from miles around. Vast quantities were taken upriver to Dolgellau in the eighteenth century. But though so tampered with, Arthog bog is still a very interesting botanical locality, with vegetation varying from that typical of very acid peat in the centre, to those plants which flourish in greater variety in less exacting conditions round the margins. Two interesting species here are the wavy St. John's wort, which, as Peter Benoit and Mary Richards point out in their *Flora*, is here at 'the northern limit of its known world distribution'; and the greater spearwort, a fine, large-flowered, aquatic buttercup unknown elsewhere in Snowdonia. Sedge warblers, reed buntings and grasshopper warblers breed in the thick cover of the marginal ditches and there is usually a small colony of redpolls in the birches.

Anyone interested in the vagaries of coastal formations could put in a lifetime of study on the shores of Cardigan Bay. Between Dysynni and Mawddach there are soft, eroding cliffs, which are for long stretches merely of boulder clay as at Llangelynnin. The rapid wearing away of such cliffs has its consequence farther up the coast in the formation of shingle banks, and of the massive dunelands of Mochras and Harlech, for the drift of waste material carried by the sea on this coast is to the north because of the strong and persistent south-west winds. This northward piling-up of material is obvious when you see how the south shores of the estuaries of Dovey, Mawddach and Dwyryd each have on their south sides a north-pointing extension almost closing the river-mouths, which are consequently shaped like the mouth of an old male salmon with a hook-shaped lower jaw. The shingles and dunes of the Mawddach at Fairbourne have not only built up northwards but even turn eastwards up the estuary—an unusual and puzzling development. There are two other noteworthy features of this coast. First, the long shingle banks which, instead of lying along the shore, extend straight out to sea at right angles to the coast. There is a short one off the Dysynni and another off Mochras that extends under the sea for fourteen miles and was a dreaded source of danger in the days when Cardigan Bay was full of ships. These shingle banks, though natural, have long been 'explained' in legend and story as the remains of ancient embankments which formerly kept the sea out of a low-lying region called Cantref y Gwaelod now lost beneath the sea. What is of interest is that much land here has undoubtedly been lost to the sea, as is proved by the large quantities of peat and tree-roots that lie hidden under the sandy beaches and are in places exposed on the shore. These trees—they are pine, birch, alder and oak—probably flourished four or five thousand years ago. Their remains on the shore no doubt puzzled the people of the Middle Ages and, together with the mysterious seaward-stretching shingle banks, no doubt inspired the ingenious story of Cantref y Gwaelod. Not that modern geographers seem to have much more idea than the old story-tellers had of how these shingle banks came into being and have kept their shape down the centuries. Leland (1536) was one of the earliest to record the ancient trees: 'At low water about all the shores of Aberdovey and Towyn in Merioneth appear like roots of trees . . . where now the wild sea is, were once two commotes of good plentiful but low land now clean eaten away.'

From the Mawddach estuary north to the Vale of Ffestiniog the Rhinog range climbs and dips over fifteen miles of the roughest country in Snowdonia. Its hard Cambrian grits evidently favour the growth of heather above all else and this coarse beard of heather, together with the broken nature of the rock, have understandably made the Rhinog unpopular walking country. But if you are prepared to risk your ankles in a wilderness of boulders and deep fissures dangerously masked by heather and battle your way up to the tops, you will be rewarded by finding yourself in a high land strangely different from anywhere else in North Wales. The north end of the range rises to a flattish ridge where there are remarkable tables of utterly bare rock cracked open like the mud of a dry pond. Or you wade and stumble through waist-deep heather to the shores of small, clear lakes like Llyn Pryfed and Llyn Twr Glas, lying in basins of rock on top of the watershed. Farther south on Rhinog Fawr, Rhinog Fach, Llethr and Diffwys, the going is easier: they may have shattered craggy sides but their tops are free of heather and good to walk upon.

With all the luxurious mantle of common heather on the tops and some wonderful stands of bell heather down the slopes, I need hardly say that there is great acidity in most of the Rhinog soils. But even here some of the crags are not wholly lime-deficient. In the several rivulets that cut through the screes on the east face of Diffwys there are quantities of mossy saxifrage and rose-root; and a fair amount of bladder fern and lesser meadow-rue. But the rock is perhaps not calcareous enough for mountain sorrel or green spleenwort. Alpine scurvy-grass, vernal sandwort and even moss campion were recorded from Diffwys in the nineteenth century and may possibly be there. Mossy saxifrage grows also on Rhinog Fach, the rocks by Llyn Bodlyn, and in Cwm Moch above Trawsfynydd Reservoir.

Though the eastern approaches of the Rhinog consist mainly of sour-looking moorland, there are at least pockets of base-rich soil, for one of the most crowded beds of globe flowers I ever saw was at the northern end of this moor. On the west side of the range above Harlech, where the slopes are drier and better drained, are very fine pastures coloured with saw-wort and the upright vetch (*Vicia orobus*). I have also seen moonwort there and a fine display of fragrant orchids. The birds that come to my mind when I recall my days on the Rhinog are mostly the heather birds: the deep-voiced grouse, the piping ring ouzel, the little dark merlins that have slipped quietly by on sharp, flickering wings; and of course the countless pipits. The least likely bird I ever saw here was a hoopoe on the hills above Llanbedr. This was a bird of remarkably regular habits. I was told that it roosted in a certain wood and that if I waited at 8.30 in the morning I would see it fly out of the trees to the top of a spruce a quarter of a mile away, that it would then call *oop-oop* a few times and fly towards the Rhinog. When next morning I stationed myself as instructed, everything happened just like that! This well-regulated hoopoe kept to its careful schedule for two or three weeks and then, as hoopoes always do, it disappeared as abruptly as it had come. I daresay many, if not most, wild birds keep to a similar regular time-table whenever they can but we just do not concentrate our attention on one bird long enough to find out. We often speak of wandering parties of tits but it would not surprise me to learn that their movements through the woods are in fact regular and deliberate rather than erratic.

This land above Llanbedr is an altogether delectable reach of country, a wide shelf at a height of 300–500 feet, very thinly populated by a scatter of farms and cottages, a land of small green fields and boulder walls and patches of wet-floored woodland; the gleaming sea always in front of you and the bare ramparts of the mountains behind. It is not surprising that ancient man settled here. South of the Ysgethin stream his remains are even more copious. Here the scene is higher and wilder than north of the Ysgethin; there are no small fields, only rock-scattered moorland criss-crossed by straight lines of walls. One flat stretch of several acres is a complete sea of rocks, a grey waste of boulders lying not on earth but on other boulders, and among them countless holes and cavities going down to unknown depths. All this riot of free-lying stone inevitably suggested to early man that here if anywhere was the place to set up his cromlechs. So here are the great chambered tombs of Carneddau Hengwm; and several other cairns and dolmens no doubt long since despoiled of most of their stones. And close to them all is the lake called Irddyn, not surprisingly a water with legendary associations, especially lying where it does amid a chaos of fallen rock under mysterious-looking shadowy cliffs. Close to Carneddau Hengwm is the well-constructed hill-fort of Pen Dinas, which looks across to an even finer one still with its thick foundation walls intact on the top of Craig y Ddinas, a sharp spur on the side of Moelfre, a fine 2,000-foot dome that looks over all this rushy plain, all these four thousand years of human history.

Two lakes lie higher up than Irddyn. The first is Bodlyn, deep-looking and cold, under vertical, north-facing cliffs and with much bog and sharp-flowered rush around its edge. In this lake is that rare fish the char, called in Welsh the *torgoch* (red belly) from the colour of the males in November when they come into the shallows to breed. Most of the rest of the year they keep to the depths, for they are fish of the northern world and cold water is vital to their existence. An angler at Bodlyn told me that there are also char in lakes Irddyn and Ybi, but anglers are not famous for veracity and I cannot vouch for the statement. Bodlyn's inlet stream comes down from a smaller, shallower, stony-edged pool called Dulyn, which much resembles Aran and Gafr, the lakes under Cader, for like them it has screes above and moraines all round and has a variety of plants including the common quillwort, the water starwort (*Callitriche intermedia*) and carpets of lobelia and shoreweed.

Diffwys and the ridge along to Llethr show a striking contrast between one flank and the other. On the west, steep smooth grassland going down towards the sea; on the east, the typical ankle-spraining heathery rocks of Rhinog going down terrace after terrace to Cwm Mynach —the Monk's Valley, a memory of Cymer Abbey and its sheepwalks. Here among the heather is a shallow lake rich in aquatic plants where yellow and white water-lilies float side by side. The rocks above the lake are wet and acid and the heather grows among tussocks of bog moss, the sort of place to look for the lesser twayblade, an orchid which is admittedly small and insignificant to look at, yet an exciting find anywhere on the Welsh mountains and has lately been found in quantity on the Rhinog.

I have mentioned several of the Rhinog lakes but there are over a dozen others. All are small and most are remotely placed among the rocks and heather with never a road within miles. Llyn Tecwyn Isaf stands apart in being a semi-lowland pool with a woodland background, a beautiful pool with moorhens, dabchicks, water-lilies and floating mats of marsh St. John's wort round the edge. From here you can climb a thousand feet up the slopes to the south-east to a bare-margined tarn called Caerwych, which looks across to the little black-shadowed lake of Eiddew Bach, from where it is only half a mile gently down to the much larger Eiddew Mawr. This is quite a botanically rewarding locality. Mary Richards and I found Tunbridge filmy fern deep in a vertical crack near the lake. In the lake grows a small water-lily which we hoped might be the lesser white water-lily (*Nymphaea occi-*

dentalis), a Scottish and Irish species which would have been a first record for Wales could we have substantiated it. But the road to botanical glory is beset with snares. When we sent our specimens to two esteemed authorities, one replied: 'Yes, decidedly *occidentalis*'; but the other said: 'No, it is only a dwarf mountain form of the common white water-lily.' So much for expert knowledge.

In the heart of the Rhinog are a handful of lakes lying in country that seemed 'most dismally rocky and barren' to John Loveday who saw it in 1732, and it would be the same I suppose for many people today. They are lakes set among crags and screes and a wilderness of scattered boulders. Into some of them, such as Llyn Hywel, smooth and slabby mountainsides plunge straight into deep water down slopes so bare and planed they might have been left by the glaciers this century instead of ten thousand years ago. Around these lakes the skeletons of the mountains are naked and the bedding of the Cambrian rocks is clear for all to see. Gloywlyn, Morynion, Perfeddau, Cwm Hosan, Hywel, Ybi, Du and Dulyn: all these lakes lie close under the backbone of the Rhinog. Long may they remain, unexploited, tranquil and remote, for the spirit of man needs such retreats.

Short, delightful streams hurry down off Rhinog. There are Crawcwellt and Gamlan draining the wet eastern moors and plantations; Ysgethin, Nantcol and Artro on the west. The last two are best known and have roads along them, historic houses and a long human story. But they are still rather wild and beautiful and have some notable single-arched bridges. The naturalist will find several interesting bogs in the Nantcol valley, at least one of which is a haunt of the marsh fritillary, a butterfly commoner on the coastal bogs than it is on the hills. I have never seen the hay-scented buckler fern nor Tunbridge filmy fern in the Nantcol or Artro valleys but both are recorded recently from there. The touristy Artro, famous for its sewin, has beautiful woodland along it, and, near Coed Crafnant, some remarkable long spines of rock that stand up not on the slopes where outcrops are to be expected but in the flat bottom of the valley among the peat bogs.

These spines exhibit perfectly the strata and dip of the rocks as if they had been thoughtfully arranged for the benefit of geology students. For botanists there is the long-leaved sundew to be found in these bogs, along with many other peat-loving plants.

I said there were roads up Nantcol and Artro but, mercifully, both lead to dead ends. No modern road with its stinking, rowdy engines crosses the Rhinog range. Only the old walking and riding roads are there, and if you enjoy puzzling out old ways, here on the Rhinog is ample scope for your talent. For here are drovers' roads, medieval trading roads, mine roads, peat roads and doubtless, if we could identify them, the roads of ancient man. I mentioned the copious traces of prehistory south of the Ysgethin. But dolmens, standing stones, circles and tumuli continue up the whole west side of the Rhinog range as far as the Dwyryd. Most are charted in the Merioneth *Inventory of Ancient Monuments* but quite a number that were known to earlier antiquarians such as Lhuyd have been lost and may yet be discovered by keen searching; for though some will have quite disappeared, others are probably quite visible and have been lost only because of the inadequate directions left by their discoverers of two or three centuries ago. The fact that in Cardiganshire only a few years ago a large, well-constructed Roman fort was brought to light in a most frequented lowland area is sufficient hint of the possibility of further discoveries in wild country like the uplands of Merioneth.

The old Rhinog roads mostly radiate from Harlech or Llanbedr, the ancient ports of Ardudwy. There is a fine one from Llanbedr by way of Eisingrug, Llandecwyn church, through the rocky defile by Llyn Tecwyn Uchaf to the Vale of Ffestiniog. Another branches off this and goes by the well-ridged hill-fort of Moel Goedog, which means the Wooded Hill, and so is yet another place-name reminder of trees since gone. From here the track climbs across miles of moor and bog to Bryn Cader Faner and then down to Trawsfynydd or Ffestiniog. Both these roads are significantly routed by way of standing stones, circles and tumuli and are probably ancient. But the best of all

Rhinog roads goes up from Llyn Cwmbychan and squeezes through the pass on the north side of Rhinog Fawr by way of the celebrated Roman Steps. This road begins as a narrow track winding up from the lake through a birch and oakwood lively in spring with willow warblers, redstarts and pied flycatchers. It makes towards a deep gash in the forbidding, bare mountainside in front of you. This is the true Rhinog country, these ramparts of rock too soil-less and dry for heather or any other plant, naked shelves tiered one above the other in a manner quite unlike the rest of Snowdonia. The defile the road goes through is a gorge choked with boulders and scree, an unpromising route for road-making but the only one available. So the steps were built. Although, because of the proximity of a Roman camp, they were inevitably called Roman, these steps are undoubtedly medieval and were probably built in the fourteenth century to effect a swift pony-track between the two most important places then in Merioneth: Harlech and Bala. Not only was there a need for political communications (Pennant tells us that Bala 'seems to have been dependent on the castle at Harlech'); there was also the considerable trade in wool and other merchandise between Bala and the coast. These steps are very well preserved: at first they are scores of narrow slabs about thirty inches wide laid alongside each other, climbing as an inclined path; and then becoming larger, squarer blocks and going up the pass as real steps, picking an astonishingly direct course through the waste of scree and loose rocks, a course excellently surveyed and cunningly gradiented, each step buttressed on either side with carefully sited squared stones. With little repair these hundreds of steps have kept together through the wear and weather of centuries and still take you as easily as ever right to the top of the pass, where you get that sweeping view over the moors of central Merioneth and see where your road continues on its eastward way between Arennig and Rhobell to the Lliw valley and Bala.

It was the building of the castle by Edward I in 1286 that put Harlech on the map. In those days Harlech was a port, the castle itself had access to water (it has a watergate), and presumably the low sandy ground now between the castle and the sea had not yet built up. But the matter has been much discussed and disputed and adds one more to the list of problems facing the geographer on this coast. One thing certain is that Harlech's contact with shipping did not survive into modern times. So when Joseph Cradock wrote in 1777: 'There is a good harbour for ships but no ships for the harbour', he presumably meant the estuary of the Dwyryd which is three miles north of Harlech. Writers of the eighteenth and early nineteenth centuries were not kind to Harlech. Cradock's comment, 'Its houses are mean and its inhabitants uncivilised', was a remark typical of the superior English traveller. It is more pleasant to recall that the botanist Thomas Johnson, on his way down the coast nearly a century and a half earlier than Cradock, saw fit not to sneer but to make just comment on the castle, and the busy air of the place with its fair of cattle, wool and cloth.

Until about the end of the eighteenth century the road from Harlech to Maentwrog went over the high ground. You climbed that formidable hill straight up from the castle and where the road levelled out you made for the north-east with only the monuments of prehistory as your guide. The intrepid Reverend Evans did it by moonlight: 'We ascended a difficult stair-case path up the steep side of a craggy mountain and took a north-easterly direction over the trackless plain, known to our guide by several upright stones called Maeni hirion, and concentric circles of stones . . . We passed the small lake named Llyn Tecwyn Isaf . . . a little farther, environed with lofty mountains, is the fine lake of Llyn Tecwyn Uchaf . . . The moon was now rising, and her silver beams, reflected from the waters of the lake, heightened the beauty of this recluse but enchanting scene. The road is a narrow and dangerous path along the shelf of a perpendicular rock, on the left side of the lake, which is composed of shale or shivering slate, and many impending projections overhang the traveller's head and threaten him with destruction. We appeared shut in by the mountain barrier, with nothing but craggy walls of rock on either side, and before us the dismal gloom of an impenetrable forest.'

So did John Evans and many other travellers reach the fabulous Vale of Ffestiniog they had heard so much about. Only a few decades had passed since Samuel Johnson had written the very popular *Rasselas* with its theme of seeking contentment in a far-away hypothetical Happy Valley, the origin, I presume, of so many places similarly called today. So when in the 1790's you came over the rugged and inhospitable mountains into the sheltered, verdant Vale of Ffestiniog with its woods and waterfalls and its recently rebuilt hostelry, and you saw the idyllic Tan-y-bwlch estate, recently laid out with walks and vistas at great expense, you could easily persuade yourself that you really had reached the Happy Valley at last.

The Vale of Ffestiniog is still verdant, still wooded, still beautiful. The broad, slow river, its banks now invaded by the Himalayan balsam, still winds through its flat pastures to the estuary; the waterfalls and the ferns that grow in their spray are much as they have always been. But north and south of the Vale there have been such lamentable changes that one can only envy the eighteenth-century travellers for being able to visit these places before the industrialisation of Blaenau Ffestiniog in the nineteenth century and the building of Trawsfynydd nuclear power-station in the twentieth. Blaenau Ffestiniog is forgivable: it had to be where the slate was and it grew up in a planless age when it was nobody's responsibility to control undesirable industrial development. But the deliberate placing of a nuclear power-station in the heart of wild Wales and in the very centre of a National Park is surely a violent negation of everything a National Park stands for.

It is ironical that this locality was singled out in the seventeenth century for clean atmosphere and long life, as we read in the reply about Trawsfynydd that was sent in answer to Lhuyd's questionnaire: 'It hath a very healthy air and is seldom known to be without half a dozen persons aged above ninety and such as have been used to hard labour and a milk diet . . . The complexion of the inhabitants generally very clear.'

Trawsfynydd Reservoir, before the power-station came, was for thirty years a beautiful sheet of water because of its irregular shape and many inlets and peninsulas. It was evidently made on fertile soil, for trout did very well there right from the start. Very soon it attracted great crested grebes and various duck, especially tufted and pochards; and the willow-edged bogs round its shores added to its natural history interest. When for a few years before the power-station was built the reservoir was partly drained and its wide muddy shores were exposed, it became a favourite place for wading birds on passage, the wetter peat-flats being colonised by black-headed gulls. It was also a chance for enterprising plants to find a new home, one of them, the marsh yellow cress, (*Rorippa islandica*) getting into Merioneth for the first time on record. It was in those few seasons when the water was low that oyster-catchers nested at Trawsfynydd, as I mentioned in chapter 5, the first record of their doing so inland in Wales. In the moorlands around are several interesting little lakes such as Hiraethlyn which was reported to Lhuyd as 'Llyn yr Ithlyn which abounds with a very peculiar perch which hath a twist in ye tale and choice trouts'. The perch are still there but in the course of 250 years have evidently reverted to normal. What is strange is the widespread persistency of these reports of malformed fish in Snowdonian lakes since at least the twelfth century. One-eyed fish have been a particular speciality!

A mile east of the power-station is the Roman fort called Tomen y Mur, the *tomen* being the early Norman castle-mound built within the Roman walls and still a conspicuous hump in the landscape. North and south from the fort ran the Roman road, Sarn Helen, more correctly Sarn Elen, a name probably fifteen hundred years old, for *sarn* means a road, especially a paved road, and *elen* is a changed form of *y leng,* meaning 'of the legion'. To explain this mysterious 'Elen', the romancers of the Middle Ages, ignorant of its true derivation, invented a lady called Helen, British wife of a Roman emperor, and related how she was so impassioned with the idea of modernising her native Wales that she persuaded her husband to build this trunk road across the country from north to south; and the road has been named after her ever since. The full story is in the *Mabinogion* tale: *The Dream of Macsen Wledig.*

Sarn Helen can be regarded as a key feature of the National Park in the sense that it extends all the way from the northern tip to the southern and neatly divides the Park into two equal parts. And if Sarn Helen is the Park's long axis, then its short axis is the medieval road from Harlech to Bala, for this road crosses Sarn Helen at right-angles near the centre of the Park. To follow the length of Sarn Helen from the Conway to the Dovey would be a walk of several days. But Harlech to Bala can easily be done in two days: an exhilarating walk which takes you first along a rough track that twists and dips its way round little hills and woods and drops you into Cwm Bychan; then along the shining, black lake past the echoing precipices of adder-haunted Craig y Saeth; then the quick-striding section up the Roman Steps; then slower down the wide moorland and across Sarn Helen, passing the fine upstanding stone called Llech Idris and soon, appropriate enough along a medieval road, a medieval grave with a worn Latin inscription presumably incised by a monk from Cymer Abbey and telling us simply that 'Here lies Porius, a plain man'—an epitaph surely good enough for any of us. But thoughts of mortality and the brevity of life quickly dissolve as you follow the singing larks across the moors up the River Cain, over the watershed at 1,700 feet, skirting down past Siglen Las—the Green Morass, and so back to the valley of the Lliw.

Here in the Lliw you may turn over a midstream rock or two in the hope of seeing those two little fish, the miller's thumb and the stone loach, which are uncommon in Snowdonia but used to be found in this stream (see page 111). There is no reason to suppose they are not still there, for any tributary of the Dee is linked with districts farther east where these fish are common. A creature I am pretty sure you will not find is that inhabitant of lime-rich streams, the crayfish. It was crayfish that the English milord John Byng was so nostalgic about when, leaving Wales by way of Bala in 1793, he bade us his ungracious farewell: 'I would not reside in Wales on any account; the gloom of their hills and the stonyness of their roads are intolerable; the filth of the inns, the lack of straw for your horses, the leanness of the meat and the badness of their cookery, make you pay dearly for a view and for the sight of a waterfall. Then I turn my look homeward; and sigh for a flat gravelly or sandy ride in Bedfordshire, to eat good bread and to sup upon cray-fish . . .'

CHAPTER TEN

Berwyn, Aran and Cader Idris

This southern section of the Park begins on the heather-moors of the Berwyn, extends just over thirty miles south-west and ends on the south Merioneth coast. It is bounded in the north by the line Bala-Dolgellau-Fairbourne and on the south by a line from near Lake Vyrnwy (which is not quite in the Park) by way of Dinas Mawddwy, Corris and Machynlleth to Aberdovey.

If you follow the Welshpool to Bala road you enter the National Park where the Berwyn grouse moors begin near Milltir Gerrig. From here westwards lies a fine wild stretch of heathery moorland climbing here and there to over 2,000 feet and extending to Foel y Geifr six miles away. Pennant spoke of 'multitudes of red grouse and a few black' on these moors in the 1770's. Just a century later we get a similar comment from an anonymous traveller who reports: 'The Berwyn hills are wonderful hunting grounds for grouse. Six or seven hundred birds have fallen on a twelfth of August.' Now, after nearly another century, though grouse are said to have generally declined through disease there are still enough there to warrant the employment of keepers. As for black grouse, they seem to have remained very few ever since Pennant's day and may even have died out and been re-introduced. But whatever have been their past fortunes in this area, they are now increasing in the forestry plantations at the head of Cwm Pennant a few hundred yards from the Park boundary.

The modern road from Milltir Gerrig to Bala is rather circuitous. But there is an ancient track which takes a more single-minded route across country, its general direction being straight from near Milltir Gerrig to Bala by way of Maes Hir and Plas Rhiwaedog. Any plant-seeker taking that track should look out for three rarities which, though as far as I know are not recorded for the Park section of the Berwyn, are found elsewhere on the range. They are bog rosemary, which is particularly rare as a mountain plant; lesser twayblade, long known just outside this edge of the Park; and cloudberry, which grows abundantly on the highest summits of the Berwyn four miles east of this boundary of the Park and is not known elsewhere in Wales. The west side of this moorland can be explored from Bala by way of the Hirnant valley. Down there in the hedges near Rhos-y-gwaliau in May you can see the lovely flowers of the bird-cherry, which is by no means a common tree in most of Snowdonia. Another local species, the stone bramble, grows on the rocky, well shaded banks of the Hirnant stream; and nearby in a damp streamside field is a colony of the wood horsetail, a species which, though typical of western and northern Britain, is not at all frequent in the Park.

Bala Lake, by far the largest natural lake in Wales, famous for its beauty, its legends, its floods and its fish, was visited and described by many of the early travellers. Most of them more or less repeated what Pennant had said, and little wonder, for at Bala we have Pennant at his best: precise, unemotional, informative. 'Bala Lake', he says, 'is a fine expanse of water near four miles long and twelve hundred yards broad at the widest place: the deepest part is opposite Bryn Goleu where it is forty-six yards deep, with three yards of mud; the shores gravelly; the boundaries are easy slopes, well cultivated, and varied with woods. In stormy weather its billows run very high . . . It rises sometimes nine feet and rain and winds jointly contribute to make it overflow the fair vale of Edeirnion . . . Its fish are pike, perch, trout, a few roach, and abundance of eels; and shoals of that Alpine fish, the Gwyniad, which spawn in December and are taken in great numbers in spring or summer. Pike have been caught here of twenty-five pounds weight, a trout of twenty-two, a perch of ten and a gwyniad of five.'

Since the 1770's three things have changed. First, a new fish has evidently been added to the lake, the grayling, for it is now abundant there and is said to have been deliberately introduced. It is curious that the grayling, normally a fish of swift rivers, should thrive in this deep water. Second, the gwyniad, having lost favour as a food-fish, is no longer netted except by zoologists who are studying this unique species. Third, the lake is no longer such a threat to 'the fair vale of Edeirnion' since its level was lowered a few years ago and its exit-stream is more under control. As for the perch of ten pounds, this would be a British record if it could be established. In this connection Sir Herbert Maxwell's comment is perhaps worth recalling though it may undermine our faith in Pennant: 'Pennant was a good naturalist, no doubt; but he was even more renowned as a traveller, whose business it was to make his tales readable. He may have refrained, therefore, from making due allowance for that remarkable property in sporting fish which causes them to increase continuously in weight after death.'

If Bala Lake's fish are interesting its birds are less exciting. Lakes with little vegetation to provide cover are rather useless as breeding places for waterfowl and in winter this lake and the country round evidently offer little in the way of food for surface-feeding ducks. Nor are diving ducks very numerous. Summing up the ducks as a whole we can say that mallard, pochard, tufted duck and goldeneye are the most regular species. But recorded observations by bird-watchers on and around this lake are regrettably few and as in so many places in Snowdonia, an increase in observers would undoubtedly show that Bala Lake attracts more birds than records suggest.

In the winter of 1783–4 'numbers of ducks were caught frozen to the surface', reports the Honourable John Byng, who also remarked on the absence of mute swans: 'The lake belongs to Sir Watkin Williams Wynn who has for certain reasons (unaccountable) taken away the swans, one of its beauties.' Presumably Sir Watkin who, as Pennant put it, 'claims the whole fishing of this noble lake', had decided that swans and fisheries did not go well together.

Three other large birds visit this lake: whooper swans

irregularly, usually in hard weather; cormorants regularly, especially immatures; and of course herons. The yellow wagtail, formerly widespread but very thinly distributed as a breeding species in parts of Snowdonia, has lost ground in Britain this century. It used to nest about Bala Lake and, assuming its range has shrunk away south-eastwards, this would now be its most likely haunt in the Park. But I have made several vain searches in the likely-looking wet fields and pastures near the lake and I fear that this attractive bird is lost as a breeding species there and throughout Snowdonia. Precisely the same can be said of the red-backed shrike which 50 years ago was also a frequent nesting species around Bala Lake but has unaccountably declined along with the yellow wagtail. Among the species still breeding at the head of the lake are mallard, curlew, common sandpiper, whinchat, sedge warbler, reed bunting, meadow pipit, skylark; and a few sandmartins which nest in the clayey banks of the Dee.

The north-country naturalist George Bolam, who settled at Llanuwchllyn near the head of Bala Lake for a few years at the beginning of this century, wrote an informative book called *Wild Life in Wales* which is mainly about the Bala country. It comes strongly from his pages that wild life was more abundant then than now and one feels more than a tinge of sadness at the thought of species now gone or going through human agency. I am not suggesting that the disappearance of the yellow wagtail or the red-backed shrike is anything but part of a natural cycle that may ultimately turn again in their favour. But there is so much more pressure now on the habitats of wild life; agriculture has become so much more aggressive with machines and chemicals that wild life is always steadily retreating even in the wilder parts of Snowdonia. But for practically all species the reduction in numbers cannot be proved because their populations were never censused in the past.

The lakeside plants at Bala are worth looking at. The lowering of the water-level has left a stony beach several yards wide which was quickly colonised by docks and other coarse weeds but perhaps these are only an early stage in an interesting succession. It might also be worth looking here for those chance aliens so beloved of some botanists. That attractive goosefoot, Good King Henry, has certainly been there a long time but there may well be other adventurous species, especially as there is a railway, now disused, alongside the lake and you never know what strange seeds may have been accidentally brought there by past trains. The south-east shore has also long been known as a site for the lesser meadow-rue (*Thalictrum minus* subspecies *umbrosum*) here in a typical shady, lake-edge habitat. It was known there in the seventeenth century, being reported to Edward Lhuyd in answer to his questionnaire. Some of the hedges in the Bala-Ffestiniog district intrigue visitors not accustomed to seeing the pale-pink spikes of the willow-leaved spiraea as a countryside plant. Introduced by a local landowner last century, it has been used in the hedges over a wide area.

On the slopes that rise from the southern shores of Bala Lake there are more ash trees than commonly grow on Snowdonian foothills. This is a sure confirmation of that lime-richness of soil I have mentioned in chapter 2 as being characteristic of this side of the lake. The farm name, Maes Meillion, which means 'clover-field', is also probably significant of this extra fertility, for clover thrives best on the better soils. On these slopes you can find meadows and sheep pastures nearly as fragrant and colourful as meadows in the limestone Alps. In the wetter places there are marsh orchids. On drier fields there is a scattering of frog orchids and also—but it is very rare and hard to find—the small white orchid, which characteristically shares these calcareous pastures with the frog orchid. In a wet alderwood sloping to a stream the two species of spotted orchid are unusually close neighbours along with an abundance of the marsh hawksbeard, here close to its southernmost limit in Britain. But many people would say that the glory of these slopes is the upright vetch (*Vicia orobus*) that makes splendid patches of purple-pink in some of the fields.

Aran Benllyn, the peak you see from Bala, towering over the western end of the lake, looks even finer the nearer you get to it. Go, say, to Talardd in Cwm Cynllwyd and you begin to get the details of this crag-topped mountain and its exciting perpendicular cliffs plunging out of sight into a deep hollow. If you are a botanist you will not fail to observe that these cliffs are east-facing and so will have gullies that even in a hot summer will be cool and damp enough to keep their plants alive. Though this is yet another botanical site discovered as long ago as Edward Lhuyd's time, it has undoubtedly had less attention from plant-seekers than many other Snowdonian crags, and good finds may be made there yet. Lhuyd himself was enthusiastic about it, far more enthusiastic, in fact, than the records justify. He writes to his cousin David Lloyd in 1686: 'Aran Benllyn is I hear too far from you, else I am sure you might find there twice as many plants as on [the Berwyn]. Divers gentlemen have gone from London, Oxford and Cambridge to Snowdon, Cader Idris and Plinlimmon in search of plants; but I find there were never any at Aran Benllyn: the reason I suppose may be because it is not so famous for height as the fore-mentioned hills, but to my knowledge it produces as many rarities as Cader Idris.' The plants found by the youthful Lhuyd on Aran Benllyn's east cliffs on his first visit, on 17 April 1682, were 'by ye rivulets that run through ye rocks above Llyn Llymbran'. The name he gives the lake is interesting, for although modern maps give it as 'Lliwbran' you can still hear it pronounced locally as Lhuyd spelt it. The map version would therefore seem to be incorrect. 'Llymbran' naturally suggests a corruption of 'Llyn Bran', which might well be the original name, for Bran was an important figure in early Celtic folklore. It is also true that *bran* is Welsh for crow and that 'Llyn Bran' could be the Lake of the Crow: but this seems less likely than a link with a character from the *Mabinogion*. But 'Llymbran' may well be only a corruption of some ancient word that had nothing to do with either *llyn* or *bran*. Some authorities give it as 'Llymbren'.

Another incidental point about Lhuyd's visit to Aran Benllyn was that near Llanuwchllyn he found what is in Wales a very rare plant, the spignel, though in his time it may not have been so rare, for in those days it was esteemed for its medicinal value and may have been cultivated in herb gardens. He says he found it in a field called Bryn y Ffenigl, which translates as The Hill of the Fennel. But since fennel is almost inconceivable on an inland Welsh hillside, and since spignel would grow in just such a site, and since both are fine-leaved unbellifers, it seems certain that there had been a confusion of names here and that this name really meant The Hill of the Spignel. After all, spignel is a long-enduring plant, making great patches, and would be quite conspicuous and enduring enough to have a field named after it. I should add that as a result of my mentioning these facts in the first impression of this book, an enterprising reader has traced Lhuyd's locality and found the spignel still growing there—a nice example of the long duration of plants in upland districts where field usage has changed little over several centuries.

Most people go up to Aran Benllyn not by way of Llyn Lliwbran but by crossing the old railway at Llys and then working up the ridge. This goes up in a series of large steps which repeatedly lure you into thinking you are just under the summit only to show you another great buttress rising ahead. But there are compensations, for every higher step you reach gives you a finer view as you raise the mountains all along the horizon: the rounded shape of Rhobell Fawr, the black cliffs of Dduallt, the twin tops of Arennig, the Denbighshire hills, and so round to Berwyn, with an excellent view of the Bala fault-line and the lake lying along it. Then from nearer the top the heights of Merioneth are dwarfed as the Caernarvonshire mountains come peaking up far away. Berwyn, from the top of Aran, appears as a succession of undulating grassy or healthy hills, ridge beyond treeless ridge away to the eastern horizon. It recalls the view south from Plynlimon, as it should, for both landscapes have been fashioned out of rocks of the Silurian age. So when you look east from Aran to Berwyn you are looking at rocks millions of years younger than when you look west from Aran to Cader Idris, Rhobell Fawr, Arennig, Snowdon and the rest of Caernarvonshire.

There is no far view south from Benllyn as you pick your way among rocks and peaty hollows a mile or so to the slightly higher top of Aran Fawddwy. Once there, you have a great sweeping prospect over parts of Cardiganshire, Montgomeryshire, Shropshire, Radnorshire and beyond to the younger rocks of the Black Mountains and the Brecon Beacons 65 miles away. The Beacons may remind us that there are no Old Red Sandstone rocks in Snowdonia. Just as Benllyn looks down its crags at Llyn Lliwbran, so Aran Fawddwy drops its precipices to Llyn Craiglyn Dyfi, the source of the Dovey river. All these glorious four miles of east-looking heathery cliffs are worth exploring. There are good mountain plants such as the green spleenwort, and that rather choosy calcicole always gives the bòtanist hopes of other good things. Then there are ravens and ring ouzels, kestrels and buzzards. And it was on the slopes round the Dovey lake that George Bolam, while botanising on the cliffs above, saw a pine-marten and watched it hunting for half an hour. He also found 'a very good flint arrowhead' in a nearby stream-bed, good evidence that man had affairs to attend to in those remote corries maybe three or four thousand years ago. The lake itself has marvellous legends told about it and so also has Llaethnant, the name of the Dovey valley immediately below the lake. A strange name this, for on the face of it it means the Valley of Milk, and there is a legend to account for it. But like so many place-names 'explained' by legends, this may well be a corruption of some name now lost, a name that had no more to do with milk than Llymbran probably had to do with *llyn or bran*. From Aran Fawddwy, instead of descending into Llaethnant, though this is a fine valley of great natural history interest, you can go south into the delectable valley of the Cowarch. But do not go the obvious way, down Hengwm, unless you want a dull two-mile descent through a jungle of bracken. Choose instead the Camddwr stream and go down with its slides and water-falls or you will miss the best of the Cowarch valley. The upper part of this valley with its great crags and scree slopes, its mineral-rich rocks, its gorges and cascades, and its remoteness from internal combustion engines, is a place that deserves to be set aside and safeguarded from every threat of change or exploitation. Two and a half miles down this valley you reach the Dovey. Then it is but a step to Dinas Mawddwy.

It must be admitted that to arrive at Dinas Mawddwy in the middle of a chapter is not very appropriate, for Dinas Mawddwy is really a gateway to the Park. An excellent gateway too, because it brings you so dramatically into the mountain region. Maybe thirty miles lie behind you from where you crossed the border between England and Wales, yet it is only now, as greater hills close suddenly in on you, that you are really in country that looks essentially different from much of east Wales or even the marches. Here in the land of Mawddwy the Dovey valley narrows abruptly, its sides become steeper and its tributaries come leaping over the rim of the moors and drop down very steep dingles and waterfalls to the main river. In the fertile bottom of the narrow valley are squeezed a few fields and hedges and an abundance of deciduous trees that make this one of the most perfect valleys of Snowdonia, whether you look up it from Dinas Mawddwy or down it from Bwlch y Groes. Along it you will find a hamlet or two that look more eighteenth-century than twentieth and a scatter of delightful sheep-farms lying under the grassy slopes that are the domain of the sheep. In this valley, as in others folded inside the Welsh mountains, you get the feeling that whatever modernisation goes on in the world outside, here the old ways and the old language will endure a long while yet. Some of the farmers may own modern cars and be travelled and world-minded in a way that would have astonished their grandparents. But meet them on the mountains and they talk best about the enduring things around them: the sheep, the grass, the soils, the rocks, the quarries, the mines, the old houses and the ancient roads of their locality. They like to share with you the snippets of local knowledge they got from their fathers and grandfathers. They are born historians, handing on history by word of mouth but alas, seldom putting a word of it on paper.

Past travellers were often struck by the poverty of such

places as Dinas Mawddwy. In 1732 John Loveday, though he found some slate houses, saw others built of 'hurdles and mud and thatched with fern, with turf on ye ridge of ye roof'. Such walls have not entirely disappeared yet. Down at Machynlleth, he found that the church 'was never paved, so every Sunday ye clerk gets clean rushes to spread over ye floor, which makes a church not unlike a stable'. Sixty-one years later the poverty of the houses and the isolation of the people were what impressed William Hutton: 'Although in England I appeared like other men, yet at Dinas Mawddwy I stood single. The people eyed me as a phenomenon, with countenances mixed with fear and inquiry. Perhaps they took me for an inspector of taxes; they could not take me for a window-peeper, for there were scarcely any to peep at and the few I saw were in that shattered state as proved there was no glazier in the place. Many houses were totally without glass.' Despite all this wretchedness these travellers were amazed when they got to know them to find the people cheerful and hospitable, that none begged and that none were in rags.

The Gribin, which rises steeply to nearly 1,900 feet on the north-west side of Dinas Mawddwy, is a hill worth climbing for the view it gives of the shaping of the land. From there you look east across what was once a high plateau that has since been deeply carved by many streams. All that is now left of the old plateau is a succession of ridges all more or less at one level. Another feature of this area are the precipices round the heads of valleys, characteristic of many parts of Snowdonia. To the north, you look along the crags of Aran Fawddwy merging into those of Aran Benllyn. Near you in the north-west are the fine rocks at the head of Cwm Cowarch. And on the south-west you see the high arc of rocks on Craig Maesglasau, down which drops a very fine waterfall.

Dinas Mawddwy was a place much visited by the many fern-collectors of the nineteenth century who followed each other on the round of those Welsh crags, gorges and waterfalls most famous for specimens. Here one of their hunting-grounds, still lovely with ferns, is the shadowy gorge of the Clywedog, two miles south of Dinas Mawddwy.

Pennant knew the spot: 'After passing the Dyfi, cross a bridge over the deep and still water of the Clywedog, black as ink, passing sluggishly through a darksome chasm into open day.' The admirable old bridges that in Pennant's time carried the Caernarvon to London traffic over the Dovey and the Clywedog are still there, looking rather forlorn, for the road they were part of is gone, being superseded by the present road which is several feet higher. But the river has not changed, nor the salmon, the anglers, the poachers: all still flourish much as Pennant described. The Dovey, he said, 'abounds with salmon which are hunted in the night, by an animated but illicit chase, by spear-men who are directed to the fish by lighted wisps of straws'. The spawning salmon coming up in late autumn and early winter are still an easy prey for the poacher, especially where they squeeze up into tributaries sometimes too shallow to cover their backs. Doubtless similar stony shallows where the salmon now spawn were known to the very earliest men who hunted in these mountains.

If your ankles and your breathing are good enough there is a splendidly rough and undulating ridge walk from Dinas Mawddwy to Talyllyn Pass by way of the top of Maesglasau rocks, Craig Portas and Mynydd Caeswyn. When you get to Caeswyn you look south into a deep valley of quarries and to the slate-mining village of Aberllefenni. An interesting section of ancient road is traceable from Aberllefenni by way of the Llefenni valley to Talyllyn Pass. The Talyllyn end of it is easily visible branching off up the hillside from the present main road. In medieval times this track presumably continued the pack-horse road from Dolgellau that came up past Bwlch Coch. But a thousand years earlier than that it may have been part of Sarn Helen, the Roman road linking the forts at Trawsfynydd and Pennal. H. E. Forrest claimed to have found a Roman milestone near the highest point of this road, which would be just behind Craig y Llam, the shattered precipice that hangs over the once famous Lake of the Three Pebbles, now better known as Llyn Bach, and today mostly filled in by road construction.

If yours is not the mountain way from Bala to Dolgellau

you can have a profitable time even along the main road. For a mile north-north-east of Llanuwchllyn is the well-marked rectangle of the Roman fort of Caer Gai and from this you may follow—or attempt to follow—the Roman road westwards towards Dolgellau. This road is clear to see as a raised causeway in several places, remarkably clear when you think how easily it might have been obliterated by the plough at any time during eighteen hundred years. But as you go on it gets increasingly mixed up with the modern road and the railway but it soon recovers and takes you off at higher level down the south side of the valley to Brithdir where there is the site of a small Roman fort. If beyond there the road again gets a little problematic, why worry, now you are in Brithdir? For on this delightful 600-foot shelf with its superb view of Cader Idris and all the north Merioneth mountains you are on a terrain that will keep a naturalist happy for a very long time. Here the soil is base-rich on a broad band of limy rock that comes across from Cader Idris and goes on towards Rhobell Fawr. Here, as I have mentioned in chapter 3, there are excellent orchid meadows and there are globe flowers and moonworts; and in the Torrent Gorge the bladder fern and the hard shield fern.

There is a very delightful atmosphere about these wide shelves that look down on to the Mawddach estuary or on to the coast between Barmouth and Harlech. They are cool in summer and much of the frost rolls off them into the valleys in winter; and the views from them lift the heart. It is no wonder that ever since prehistoric time man has settled much along them. Round Brithdir in spring the hedges, thickets and the little pastures between them are delightful for their variety of birds. There are woodlarks, wheatears, whinchats, willow tits, redstarts, pied flycatchers, as well as the more usual woodland species. From Brithdir the land rises gently Cader-wards to Cross Foxes and beyond. The botanically dull acid mudstones of Ceiswyn mountain rise away on the east, but here on the wide wet tract called Tir Stent we are still on the basic rock and the flora is worth seeing. Tir Stent goes up towards Cader in a series of shallow steps, saucers of bog surrounded by rocky outcrops topped by groups of gnarled hollies, oaks and sycamores. Some of these saucers are quite acid and are rich in bogbeans, sundews, butterworts, and heath spotted orchids. Other boggy saucers are clearly less acid and have the lesser clubmoss, globe flower, and a variety of orchids and sedges. In a mile or so you reach the rough medieval pony-track that crossed from Talyllyn Pass to Dolgellau and near the old way is a field fragrant with spignel, which here is at its most southern known locality in Britain. In Britain spignel is mainly a Scottish plant, and because here under Cader Idris it is so far south of its normal range, it is under the suspicion of being a deliberate introduction though perhaps of many centuries ago. For it might have been brought there as a fodder plant. 'It has a sweetish flavour,' we read in Anne Pratt, 'reminding one of the melilot; and it is said to communicate this to milk and butter if, during spring, the cows feed upon it.' Certainly it is a delight to stand among it on this Cader pasture for the whole plant is full of fragrance. In the Alps it is a characteristic plant of high meadows and as nutriment for cattle has been esteemed from oldest times. Though the spignel looks so native and natural on its slope above Tir Stent, the proximity of the medieval road and also of the former religious house of Gwanas a mile and a half away are further indications that it was planted here for its usefulness, like that other herbalists' umbellifer, the sweet cicely, which is an obvious introduction on the roadside at Cross Foxes. Is it a coincidence that sweet cicely is also a mainly northern plant? Could not the same hand that brought it here have brought the spignel too? Above Tir Stent you are clear of trees except the few that struggle here and there among the upland rocks. Here begin the acid slopes that lead you to screes and crags higher up. On a patch of hillside bog on the mountain side of the medieval road the rare bog orchid was found in 1956 but not since. It is feared that drainage since then has now spoilt this plant's chance of survival here in almost its only recent Merioneth locality.

Talyllyn Pass is much admired by geologists as an impressive section of the biggest fault-line in Wales, a fault which extends from Bala to Towyn. Faulting often means large-scale shifting of the rocks and in Talyllyn Pass it has been shown that the south side has moved two miles north-east relative to the north side. In other words, the rocks that form the precipice of Craig y Llam and which now look across to the rocks of Geu Graig once stood opposite those above Llyn Cau. Both sides of Talyllyn Pass are of acid rocks and botanically uninviting because there is such a blanket of heather and bilberry on so many of their ledges. Craig y Llam has, I think, been very little visited by botanists and so is just the sort of place where someone some day may make an unexpected find, either a calcicole in some unsuspected streak of lime-rich rock or else some rare calcifuge. There seems no reason why the hairy greenweed, since it is found west of the pass, should not also grow on Craig y Llam, unless perhaps those north-facing rocks are too sunless for it.

Cader Idris was glowingly described by the early travellers in the language of their time. H. P. Wyndham's comment (1775): 'Few objects can be more awfully sublime than Cader Idris' can stand for all the enraptured pages written in those days about this famous mountain. We may laugh at their diction now but who can say that in two centuries from now people will not be laughing at ours? Cader Idris has come down in the world's estimation since the sixteenth century when Camden gave the readers of *Britannia* to understand that it was probably the highest mountain in Britain. It is of course not even the highest in Merioneth, being exceeded by Aran Fawddwy, and is only higher than Aran Benllyn by 26 feet. But Cader looks high from the north because it rises so steeply to its 2,927 feet from near sea-level, whereas Aran Fawddwy, merely a greater height among surrounding heights, looks less upstanding except from the deep valleys on its east. All these mountains are most imposing in hazy, east-wind, anticyclone conditions, when they are huge dim shapes looking remote and far away. Perhaps it was on such a day that Pennant described Aran Fawddwy as ' a rugged and wild summit which soars above with tremendous majesty'.

The Cader range has true mountain qualities. It has precipices dropping almost sheer to deep lakes; it has a fine high summit ridge with rocky cairns; it has great screes and bouldery slopes and high wet gullies where alpine plants are found; snow lies at least patchily on its northern face for many weeks from late autumn to spring. And until the beginning of this century mountain guides could be hired at Dolgellau inns and in earlier days from the Talyllyn side also. Some of these guides became well known through being mentioned in the books of travellers. The most celebrated seems to have been Robin Edwards of Dolgellau, who was evidently something of a character, perhaps a bit too rough in his ways and his humour to please the more genteel of his clients. But the Honourable John Byng, himself clearly a gay and original spirit, found Robin Edwards to his liking when they went up Cader together on Wednesday July 7 1784: 'Robin Edwards has shown this mountain for 40 years: and though now 60 years old is by this health-establishing custom as equal to his business as formerly: by being encouraged and laughed with by all comers, he is become a determined wit and does say many things of oddity; Mr. Cradock, in his Tour, has accused him of swearing, which angers Robin much, though Mr. C. has very good reason for the assertion.' So for a lifetime this hardy guide got his often timid clients up to the top. We are given a parting mention of him in his declining years towards 1800 by an anonymous writer who refers to 'our excellent and eccentric guide, poor old Robin Edwards'. Climbing Cader and talking about Cader all those years, what a mass of miscellaneous information Robin Edwards must have carried in his head! Another eccentric guide is mentioned in 1811 on the Talyllyn side of Cader, 'where there is a small public house kept by a most original character who in the triple capacity of publican, schoolmaster and guide to Cadair Idris, manages to keep the particles of his carcase intact'.

If the guide with the longest career on Cader was Robin Edwards in the eighteenth century, surely the most enterprising was Richard Pugh, also of Dolgellau, in the first half of the nineteenth century. For it was Pugh who built

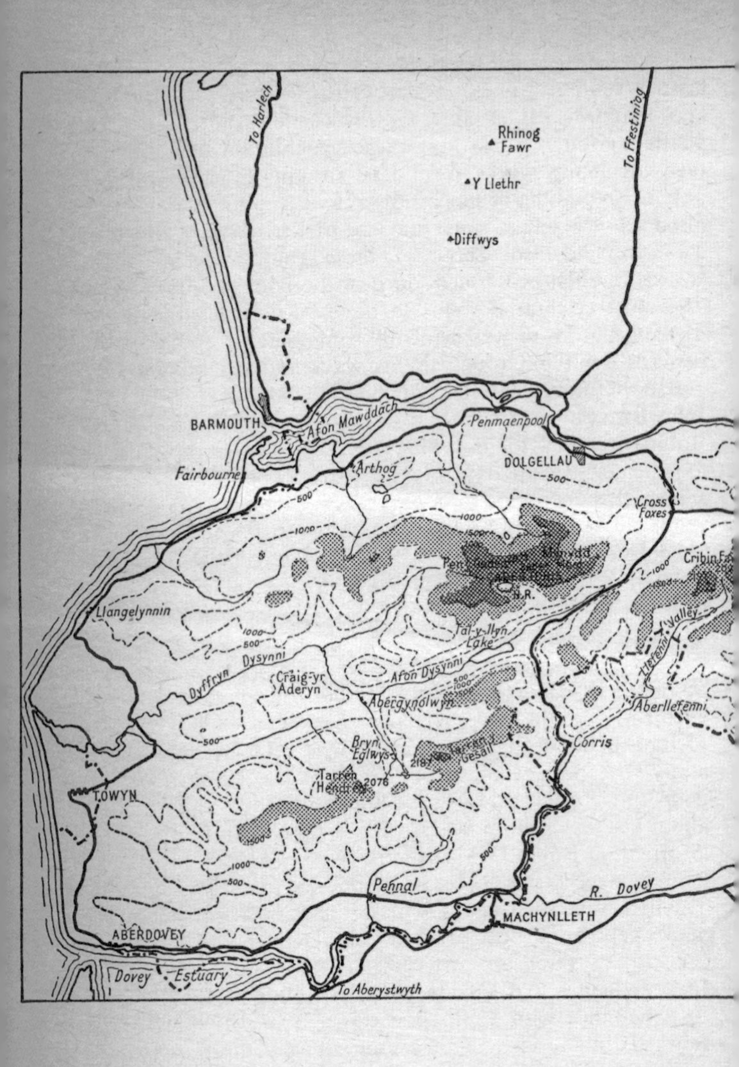

To Harlech
Rhinog Fawr
To Ffestiniog
Y Llethr
Diffwys
BARMOUTH
Afon Mawddach
Penmaenpool
DOLGELLAU
Fairbourne
Arthog
500
1000
1500
Cross Foxes
N.R.
Llangelynnin
Tal-y-llyn Lake
Dyffryn Dysynni
Afon Dysynni
Craig-yr Aderyn
Abergynolwyn
Llefenni Valley
Aberllefenni
Corris
Bryn Eglwys
2187
Tarren Gesail
Tarren Hendre
2076
TOWYN
Pennal
R. Dovey
MACHYNLLETH
ABERDOVEY
Dovey Estuary
To Aberystwyth

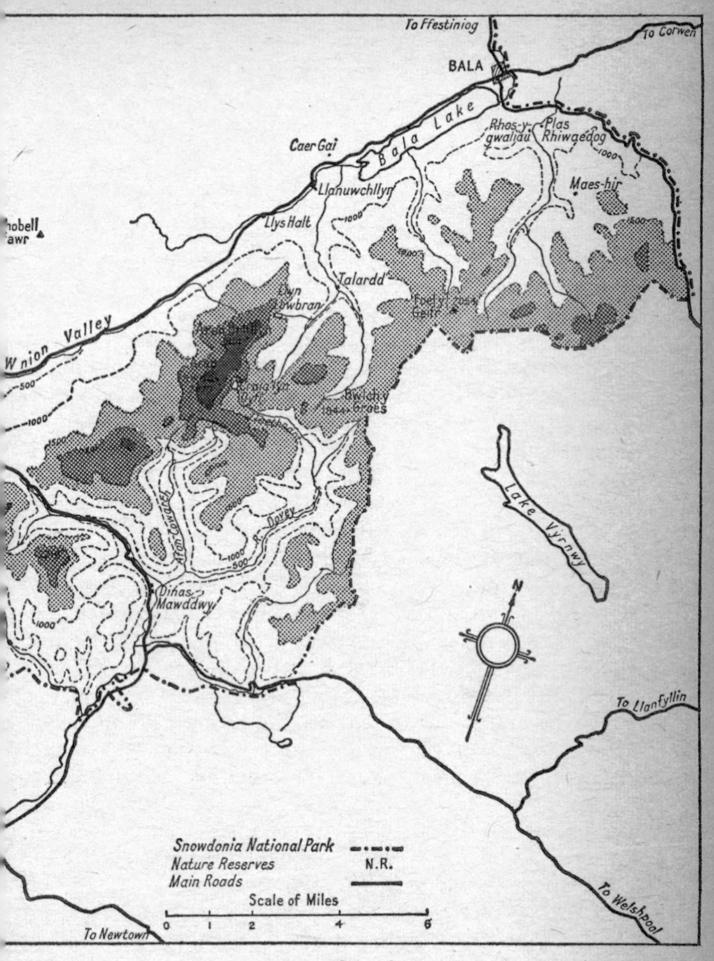

Fig. 9. Berwyn, Aran and Cader Idris

the stone hut on the summit in the 1830's. According to a contemporary description the hut was a considerable asset: 'This has proved of great advantage to visitors who before were not unfrequently assailed by the teeming shower without an opportunity of shelter; and who had no spot for temporary refreshment while waiting for the dispersion of the misty clouds in order to enjoy the exquisite prospect. Here parties and individuals may have all convenient refreshments.' Although so many thousands of words were written during the eighteenth and nineteenth centuries by travellers awestruck by the view from Snowdon and Cader Idris yet I think there is still a further point to be added. Almost without exception the climbers of those days were summer visitors and though some of them undoubtedly had splendid views from the top they might well have done better in the winter. In very severe east-wind weather, when it is hazy in the valleys, it can be exceptionally clear on the tops. In such conditions it is possible to see from Cader a vast amount of the Irish coast under snow and also the snow shining on Snaefell in the Isle of Man about 110 miles away.

In chapter 1, I spoke of the many writers who visited Snowdonia prior to 1850. The list could be continued through the second half of the nineteenth century too. Tennyson went up Cader in heavy rain in August 1856 and Francis Kilvert did the same a few years later. Kilvert's is the best description of a walk over Cader that I know. With Pugh as guide, he went up the zigzag path and came back down the loose scree of the Fox's Path. A gale and rain are not the most favourable conditions for the appreciation of mountain summits and he found the top 'the stoniest, dreariest, most desolate mountain' he ever was on. But all mountaineers know the joy he felt as he came down out of the clouds into clear weather below, though it was only short-lived: 'Down, down and out of the cloud into sunshine, all the hills below and the valleys were bathed in glorious sunshine—a wonderful and dazzling sight. Above and hanging overhead the vast black precipices towered and loomed through the clouds, and fast as we went down the mist followed faster and presently all the lovely sunny landscape was shrouded in a white winding sheet of rain. The path was all loose shale and stone and so steep that planting our alpenstocks from behind and leaning back upon them alpine fashion we glissaded with a general landslip, rush and rattle of shale and shingle down to the shore of the Foxes' Lake. The parsley fern grew in sheets of brilliant green among the grey shale.'

The zigzag path Kilvert went up has been the subject of some speculation. It is a very carefully made path up the steep grassy slope and is known in Welsh as Llwybr Cam Rhedynen—the Crooked Path through the Bracken. It has been called a Roman road even by the Ordnance Surveyors, an idea of which the Ancient Monuments surveyors were rightly scornful, for no path could look more un-Roman. But to dismiss it as they did as 'probably no more than an immemorial sheep-track' seems to be quite as unjustifiable as calling it a Roman road, for it is far too geometrical for any sheep-track. I think it most probably began as part of an old track over to Llanfihangel y Pennant and that it was gradually improved by the guides from possibly the 1790's onward and was given its present form in the 1830's by Richard Pugh when he built the shelter on the summit. Whoever the road builder was, we have a testimonial to his labour by a mountaineer of 1840: 'The road up the mountain on the Dolgellau side has lately been much improved, so as to enable ladies and gentlemen to ride to the very top with the greatest ease and safety, which cannot be done on the other side of the mountain without great danger.'

I have mentioned the high lakes shadowed under the precipices of the Cader range, three on the north side and one on the south. They are typical of the British mountain scene. But the fifth lake, at Talyllyn, is different from these. Talyllyn Lake's affinities are with Bala Lake. They both lie in the same fault-line. Both have good Welsh names regrettably unknown to most English visitors, Talyllyn Lake being Llyn Mwyngil, meaning 'the lake in the pleasant retreat', and Bala Lake being Llyn Tegid, 'the beautiful lake'—assuming that these Welsh names are not fanciful corruptions of earlier names. Both are waters in which fish thrive well, which may be connected with something else they have in common: namely, that both probably have a drainage of lime-rich water into them, Mwyngil off Cader Idris and Tegid off the slopes to the south. This may mean that both are a little less acid than most mountain lakes and therefore richer in vegetation and aquatic life. Both have certainly been famous among anglers for a very long time. Botanists visit Talyllyn mainly for its interesting hybrid water-lily. For the bird-watcher the lake is best in winter for even in the hardest weather there is usually a piece of open water where pochard, tufted ducks and whooper swans can feed. This lake is not deep and seems doomed to be divided into two at some remote future as the silt and stones brought down by the Amarch stream build out further and further towards the opposite shore.

Overlooking the flat valley of the Dysynni six miles upriver from the sea is Craig yr Aderyn, or Bird Rock, a spectacular hill that drops about 200 feet sheer on its north-west side and continues below that very steeply down grass and scree for another several hundred feet to near sea-level. This fine precipice is almost the only regular inland breeding site of the cormorant in Britain and so is a unique site worthy of the protection now given it by the West Wales Naturalists' Trust by arrangement with the owner of the property. No one knows how far back this colony dates but it is probably ancient. Lhuyd speaks of Craig yr Aderyn in 1682 and the cormorants themselves are mentioned in the 1802 edition of Camden's *Britannia*: 'Craig yr Deryn or the Rock of Birds is so called from the number of corvorants, rock pigeons and hawks that breed on it.' Many authors of guide-books have mentioned Craig yr Aderyn. Most of them have been ignorant of birds and repeated the accounts of previous writers. The 'hawks' mentioned by Camden's *Britannia* invariably get included, though hawks are certainly not an obvious feature there. Cathrall's account (1828) may be taken as typical: 'Craig Aderyn is a most picturesque and lofty rock, so called from the numerous birds which nightly retire among its crevices: the noise they make at nightfall is most hideously dissonant, and as the scenery around is extremely wild and romantic, the ideas engendered by such a clamour in the gloom of evening and in so dismal and desolate a spot are not the most soothing or agreeable. Towards twilight some large aquatic fowls from the neighbouring marsh may be seen majestically wending their way to this their place of nocturnal rest.'

Local accounts have it that the cormorant colony is at present much smaller than it was even within living memory, when hundreds of pairs bred there and both the north and west faces of the rock had many nests. But it is a curious mannerism of countryfolk everywhere to make out that natural phenomena were far more phenomenal years ago than now, and this is especially so of animal and bird numbers. Besides, the west side of Craig yr Aderyn is not precipitous enough for cormorants to nest on. Whatever past fluctuations there may have been, at present some 25–30 pairs of cormorants breed annually. As the cormorant is not on the protected list, we have at Craig yr Aderyn the unusual position of a bird sanctuary for a species that is officially outlawed. Long may that protection continue!

Followers of old roads may be interested in the local tradition that there was a pilgrims' way over the hills from Machynlleth to Towyn through the pass between Tarren Hendre and Tarren y Gesail and thence by way of Bryneglwys and Abergynolwyn. If this looks a long way round it must be remembered that the paths of those times often preferred the high ground to the forest-impeded, watery bottoms of the valleys. This old track is spoken of as a pilgrims' road to the holy island of Bardsey. But more likely, if it was a pilgrims' road at all, most of them were going only to Towyn, for Cadfan's church there was a focus of some prestige well into the Middle Ages and was itself a place of frequent pilgrimage. But as there was a strong link between Towyn and Bardsey (for Cadfan is the accepted sixth-century founder of the religious settlements at both places), doubtless many pilgrims then went on from Towyn across the sea to Bardsey. Sections of this alleged pilgrims' road are still known near Abergynolwyn as Llwybr Cadfan, that is Cadfan's Way. Whatever their origin these old paths that take you over the tops away from petrol and diesel and din are always a delight. This one might be regarded as a bird-watchers' path for it gives you a cross-section of the characteristic birds of lowland and upland on this west side of the Park.

As you climb steeply from Machynlleth on a spring day you soon lose the lowland birds for the semi-upland species. You leave chiffchaffs and blackcaps behind you in the Dovey valley but hear more redstarts, garden warblers and willow warblers. In the woodland scrub there are pied flycatchers and willow tits (but not marsh tits). Where the hill becomes open there is often a woodlark singing but at 700 feet this is really the domain of meadow pipit, skylark and wheatear. Then you enter the plantations of the Forestry Commission and the world changes abruptly. Now you are amid rapidly growing young Sitka spruces and other conifers. On the outskirts of the forest it is the songs of whinchat and tree pipit, yellowhammer and hedge sparrow that you are most likely to hear; and the weird crowing of the blackcock. You will find, too, an abundance of redpolls. But these Sitkas grow very quickly once they reach about ten feet and none of these conditions will be the same a few years hence. Four miles from Machynlleth you reach the gap in the hills at about 1,400 feet. On either side rise heather-moors with a few red grouse on their peaty tops, and ring ouzels and an occasional merlin in the heathery scars carved into the hillsides by the weathering of the ages. Once you are over the pass the track leads you down near more plantations and more redpolls, then through some spectacular old slate-quarries, the haunt of ring ouzels, kestrels, stock doves, pied wagtails, and—rather surprisingly—a few house martins nesting on the rock face. Down near sea-level you reach the wooded dingle at Dolgoch with its fine deciduous trees and waterfalls, its ravens, buzzards, dippers, grey wagtails and pied flycatchers. And here you are back among blackcaps and chiffchaffs once more.

And so to the south Merioneth coast. It is partly sands, partly shingle, partly cliffs. But when you see how small these cliffs are and how they lack bold headlands dropping into the ocean, you know why neither auks nor shags nor kittiwakes breed on them and why all Merioneth's breeding cormorants have to go inland to nest. The shingle and the sands have always been a precarious habitat for breeding birds. Only three have learnt the trick of nesting along the high-tide mark and of snatching a success some time from May to July between those high spring tides that sweep over their breeding places. But on the sandy foreshore in the last dozen years a worse threat has developed for these tide-line nesters: the increasing number of holiday-makers. Before then the beaches and dunes were not overrun by people until August when nesting was over. But now, with so many cars and caravans about, the seaside week-end habit has become established and the birds are continually frustrated from incubating by the constant presence of people who are not even aware that they have sat themselves down to sunbathe a few feet from a bird's eggs. For none of these birds makes real nests and their eggs are almost invisibly camouflaged.

The little tern's hold on this coast is extremely slender. This is one of Merioneth's rarest birds—probably an average of not more than half a dozen pairs annually—and yet it is given no special protection beyond sharing the general umbrella of the Bird Protection Acts. It survives only because the shingle on which it nests is less comfortable for picnickers' posteriors than the sand nearby. The oystercatchers are very different. They are far more numerous, assertive and adaptable. As if realising that there is no future for them on the sandy foreshore, they have crossed over the dunes and now nest in fields on the landward side, sometimes lining their nests with rabbit-droppings in default of sea-shells or pebbles. Probably the oystercatchers which are safest are those that breed on golf-courses, for the rough places of these coastal golf-courses are unintentional nature reserves. They not only harbour ground-nesting birds but also are among the few places left for bee orchids, marsh orchids, twayblades, adders' tongues and other rather choice plants. The future of the ringed plovers here seems less promising. They are perhaps too specially adapted to nesting on the seaward side of the dunes and do not look like learning from the oystercatchers. Yet there are places elsewhere, on Bardsey for instance, where the ringed plover has nested on cliff-tops, so this very attractive little bird may prove adaptable enough to survive even now.

The dunes of the southern half of the Merioneth coast are much smaller than those of the northern half and lack the wide wet hollows where botanists find so many of the rarer species, though there are lots of good plants like sea holly, houndstongue, blue fleabane, autumn lady's tresses and many others. But the essential beauty of such dunes is not in their individual plants so much as that they are such a different world from fields, hedges, woods or mountains. What makes sand-dunes so unique is their fragrant air, their clarity of light and the sheets of colour provided by such common plants as lady's bedstraw, lesser hawkbit, thyme and rest-harrow; and at every turn the long pale plumes of the marram-grass against the sky.

It remains for me to mention the extreme southern edge of the Park, where wooded hills dip steeply from 700 feet to the Dovey estuary. The hill-tops are acid moorland that is saved from botanical anonymity by providing in one bog the only known habitat in Snowdonia where all three sundews are found together. These hills are also the best woodlark haunt in the Park, for the woodlark has a southern distribution in Wales as it has in England and is hardly known in the northern two-thirds of Snowdonia.

The Dovey estuary is a feeding-place, gathering ground and fly-way for many birds. It is a meeting-place of species both from sea and land, and from north and south; and many rarities have been recorded over the years. In winter there is a variety of ducks, white-fronted geese, many waders and sometimes wild swans. In summer many species breed in the woods and marshes round about. But its outstanding birds are the shelducks, locally called lady-fowl, which on spring days often leave the estuary and stand about in little colourful groups on grassy slopes or in woodland clearings. Sometimes I have met with them along paths through forestry plantations three or four miles from the estuary and six or seven hundred feet above it, a strange place to meet with a species normally so tied to estuary or sea. They nest in holes in the ground sometimes far from the estuary. The newly-hatched young are then walked down to the estuary by their parents and many a surprised motorist has been held up along the road from Machynlleth to Aberdovey while some kind person has supervised the safe crossing of a family of shelducks on their way down from the hills. A final word about shelducks: if you do happen to find these young ones on the road please do not assume because you cannot see any adults near that the young ones have been abandoned. They are most attractive little creatures but please resist the temptation to take them home and look after them, for they are very difficult to rear as they need to be provided with special vitamins. You may be sure that one or both of their parents are hidden not far away and anxiously waiting for you to depart. So put these infants safely over the wall on the estuary side and go on your way rejoicing.

I began this second half of the book where, in the extreme north of the Park, the Roman fort of Caerhun looks down its fields to the estuary of the Conway. I end forty-five miles away in the south at Pennal where a similarly situated Roman fort guards the tidal waters of the Dovey. At Caerhun a church sits in an angle of the Roman rectangle; at Pennal there is a farm-house. At Pennal the outlines of the Roman camp have been almost completely blurred by time, but the stones of the old fortifications are everywhere. Many are in the farm buildings on the camp and you can see others in house walls in the village, in the churchyard wall and a few in the church itself. And they are scattered down the fields towards the Dovey, where a paved road now lost under the fields was clearly visible until the seventeenth century. These stones are easily recognisable for they are a warm brown granite quite different from the local blue slate. This granite, as the rector of Dolgellau reported in his answer to Lhuyd's questionnaire in 1693, came from the still-worked Tal-y-garreg quarry on the coastal hill two miles north of Towyn, and was evidently shipped to Pennal by way of sea and estuary, a distance of some twelve miles. That the

Romans should have gone to such lengths to get this especially hard rock shows the importance they attached to their Pennal camp. Such a fort, holding a key position on the Sarn Helen or Great West Road, would have had good communications. All the same, the routes they followed in any direction remain very uncertain.

In trying to follow these old ways you may or may not be successful. Perhaps it hardly matters. What seems to me important is that you will surely be led into some very good country, often far from present roads and habitations. As you discover where the Romans and perhaps peoples long before them placed their settlements and took their roads, you will be given glimpses of how this land of Snowdonia looked to people of two or three thousand years ago. As distinct from being told of such things in books you will see and feel for yourself the changes that have come about over the centuries: how valleys once clogged with alder swamp have been drained and cleared; how once forested hillsides have been turned into pastures; how uplands once populous are now deserted; how roads often went over high ground that has no road now. It is against the changing background of the past that we should view all plant life, all animal life, all human life, and I know of no better introduction to the natural history of our region than to set off along an ancient road.

Appendices

Glossary

Bibliography

Index

APPENDIX 1

NOTES ON SOME OF THE UPLAND PLANTS OF SNOWDONIA

(*excluding most of the species which are common both in lowlands and uplands*)

calc = usually growing on lime-rich or other basic rocks
C = Caernarvonshire section of the Park
M = Merioneth section of the Park. With a few exceptions the English names follow those used by Fitter and McClintock in *The Pocket Guide to Wild Flowers*

FERNS AND THEIR ALLIES

Fir clubmoss   Common
Marsh clubmoss   One C locality, a semi-upland bog. Recorded for M but site now unknown
Stagshorn clubmoss   Common
Alpine clubmoss   Common
Lesser clubmoss   Frequent. A useful indicator of less acid soils
Common quillwort   Frequent in lakes
Small quillwort   Very local in lakes
Killarney fern   Extremely rare. Most likely in dark holes behind waterfalls. No C locality now known
Tunbridge filmy fern   Rare in shady damp places
Wilson's filmy fern   Frequent on damp rocks
Parsley fern   Common on some acid screes but surprisingly absent from others
Green spleenwort   Frequent. A valuable indicator of basic rocks
Forked spleenwort   Rare. Mainly but not entirely associated with lead-mines
Bladder fern   Frequent on basic rocks and mortared walls
Oblong woodsia   Very rare in C. Extinct M? (Found Cader Idris many years ago)
Alpine woodsia   Very rare, C only
Rigid buckler fern   One old C record (Cwm Idwal)
Hay-scented buckler fern   Very rare, usually in spray of waterfalls
Hard shield fern (var. *cambricum*)   Far from frequent on basic rocks and deceptively like holly fern

Holly fern Very rare, C only
Oak fern Frequent, especially in base-rich localities
Beech fern Generally much commoner than oak fern
Pillwort In one C lake. Not now known in M
Moonwort Occasional on high ground

FLOWERING PLANTS

Juniper Local C. Very rare M
Globe flower Frequent, mainly calc
Alpine meadow-rue Frequent C calc. In M only 2–3 plants known on Cader Idris
Lesser meadow-rue Frequent calc
Welsh poppy As an alpine a rather infrequent calc in C. Rare in M
Alpine penny-cress C only. Local on Conway valley lead-mines. Not alpine
Hairy rock-cress Frequent calc
Northern rock-cress Very local calc C. Only one M locality
Narrow-leaved bitter-cress Very rare semi-upland calc M only
Twisted whitlow-grass (*Draba incana*) Rare calc C only
Alpine scurvy-grass Local C, very local M
Awlwort Rare at edges of lakes
Mountain pansy Surprisingly infrequent compared with its local abundance in central Wales
Common rockrose Rare calc of semi-uplands, e.g. N.E. of Dolgellau M. Not on mountains
Moss campion Very local calc in C. In M only 2–3 plants on Cader Idris
Alpine chickweed Extremely rare C only
Arctic chickweed Rare C only
Vernal sandwort Locally common calc mainly Snowdon and Glyder. Very rare in M except at Hermon copper-mine where common (*see* Thrift)
Hairy greenweed Very local on acid rocks of Cader Idris M only
Wood vetch C only, abundant in Cwm Glas Crafnant on basic rocks
Upright vetch (*Vicia orobus*) Locally frequent in semi-upland pastures (especially on basic soils?)
Cloudberry Unknown in National Park but as it occurs nearby on Berwyn it should be looked for on wet moorland
Stone bramble Local calc
Mountain avens Very rare calc C, not in M
Alpine cinquefoil Very rare calc C only

Lady's mantle (*Alchemilla filicaulis*) Local in base-rich grasslands
Rose-root Frequent. Mainly calc but fairly tolerant of acid conditions
Rock stonecrop Local calc of lowland and semi-upland rocks
Arctic saxifrage (*Saxifraga nivalis*) Rare C only
Starry saxifrage Plentiful on acid and basic rocks but commonest on basic
Tufted saxifrage Extremely rare C only
Mossy saxifrage Frequent calc
Purple saxifrage Locally frequent calc in C. Rare in M except on Cader Idris where very restricted
Grass of Parnassus Local C. Not M
Chickweed willow-herb Very local C only
New Zealand willow-herb Rapidly spreading, especially in semi-uplands
Alpine enchanter's nightshade Very rare
Spignal Very local in upland pastures
Wood spurge Rare on semi-upland basic rocks
Alpine bistort (*Polygonum viviparum*) Very rare C only
Mountain sorrel Frequent calc
Least willow Very local on mountain tops
Cowberry Widespread and frequent on acid moorland
Cranberry Local in or near bogs
Serrated wintergreen One locality M
Hermaphrodite crowberry Besides the very common *Empetrum nigrum*, there is also *E. hermaphroditum* which is local, e.g. Cwm Idwal and Cader Idris
Thrift Locally frequent on mountains in C. Rare as a mountain plant in M (Cader Idris). Plentiful at Hermon coppermine with vernal sandwort
Bogbean A common adornment of moorland pools
Common butterwort Frequent in bogs and on wet acid and basic rocks but commonest in less acid places
Marjoram A local lowland and semi-upland calc
Sea plantain Very local on C mountains. In M on Cader Idris only (apart from coast)
Shore-weed In most high lakes
Ivy-leaved bellflower Local in damp or wet acid places to the semi-uplands
Water lobelia Frequent in acid lakes
Northern bedstraw Very local calc, especially rare M
Slender bedstraw Very local calc C only
Moschatel Extends widely to mountains, e.g. Cwm Idwal
Coltsfoot Native by mountain streams
Mountain everlasting Widely but thinly scattered

Golden rod The compact, early-flowering mountain form is distinctive
Ox-eye daisy A conspicuous indicator of some basic rocks
Melancholy thistle Very rare. Decreased?
Alpine saw-wort Very local. In M Cader Idris only
Marsh hawksbeard Locally common in base-rich wet woods and meadows
Snowdon lily (Lloydia) Snowdonia's most distinguished arctic-alpine. A rare calc intermittently scattered in a belt 7 miles long on C mountains
Three-flowered rush Infrequent on C mountains. Not M
Lesser twayblade Very local on sphagnum-heather moors. Most frequent Rhinog and Berwyn
Bog orchid Very rare in less acid bogs. Insignificant and easily missed
Frog orchid Rare calc of semi-upland pastures
Small white orchid Very rare calc of semi-upland pastures and mountain ledges
Early purple orchid The only frequent orchid of mountain cliffs. Calc
Floating bur-reed Local in mountain lakes
White beak-sedge Local in bogs
Hair sedge (*Carex capillaris*) Very rare C only
Bog sedge (*C. limosa*) Rare in semi-upland C bogs
Tall bog sedge (*C. paupercula*) Known in only one M bog near Trawsfynydd
Slender-leaved sedge (*C. lasiocarpa*) Rare by moorland lakes C and M
Jet sedge (*C. atrata*) Very rare C only
Stiff sedge (*C. bigelowii*) Mountain tops, mainly C
Few-flowered sedge (*C. pauciflora*) One old record C
Viviparous fescue (*Festuca vivipara*) Common on high ground
Alpine meadow-grass (*Poa alpina*) Rare C. Presence in M needs confirmation
Bluish mountain meadow-grass (*P. glauca*) Rare. Not in M
Mountain meadow-grass (*P. balfourii*) Rare. Not in M
Mountain melick (*Melica nutans*) Rare lowland and semi-upland calc
Alpine hair-grass (*Deschampsia alpina*) Rare. Not in M

APPENDIX 2

A LIST OF THE BIRDS OF SNOWDONIA

In chapter 5, I have written mostly about upland birds since they are the most distinctive species of a region such as Snowdonia. Consequently many lowland and coastal species were left out. The list below aims to compensate a little for these omissions by giving notes on the status of all species. It does not, however, purport to be a complete check-list of the birds of the Park. In using this list it should be remembered that the only coastline included in the Park lies in Cardigan Bay in Merioneth.

(B—indicates a regular breeding species; S—signifies summer visitor)

Black-throated diver Very seldom recorded.
Great northern diver Probably regular winter visitor but in very small numbers. Mainly coastal.
Red-throated diver The most frequent diver, usually on the coast but sometimes on lakes.
Great crested grebe Very rare as a breeding species despite the large number of lakes. More frequent autumn to spring especially on the coast.
Red-necked grebe Very occasional visitor, mainly coastal.
Horned grebe Very occasional visitor, mainly coastal.
Black-necked grebe Very occasional visitor, mainly coastal.
Little grebe B Breeds on a few waters including moorland lakes. Commoner in winter especially on estuaries.
Leach's petrel Rare storm-driven straggler.
Storm petrel Rare storm-driven straggler.
Manx shearwater Flocks frequently off the Merioneth coast especially in spring.
Fulmar Occasional on the coast. In the last few years single birds have been seen at Craig yr Aderyn, 4 miles inland near Towyn, Merioneth, so eventual breeding there is not impossible.
Gannet Frequently visible fishing off the Merioneth coast.
Cormorant B 25–30 pairs breed 4 miles from the sea at Craig yr Aderyn, Merioneth. The colony is undoubtedly ancient.
Shag A few on the coast.

Heron B Occurs from the sea-coast to the upland bogs and lakes, but nesting colonies are few and small.

Little bittern 2 were shot on Arennig Fach about 1867.

Bittern A rare winter visitor most likely to occur in hard weather.

Spoonbill Has fairly often been seen on the Dovey estuary, but not in recent years.

Flamingo One on the Merioneth coast in autumn 1898, 4 in autumn 1913.

Mallard B The only widespread breeding duck, nesting from coast to moorlands.

Teal B Breeds very sparingly near moorland pools.

Garganey Infrequent on passage on the coast.

Gadwall Rare winter visitor to estuaries.

Wigeon Perhaps the commonest winter duck, particularly numerous on estuaries and coast.

Pintail Frequent on the coast in winter. Infrequent inland.

Shoveler Frequent on the coast in winter. Infrequent inland.

Scaup A few recorded on estuaries and lakes.

Tufted duck Widely distributed in small numbers autumn-spring.

Pochard Widely distributed in small numbers autumn-spring.

Goldeneye Widely distributed in small numbers autumn-spring. Goldeneye is commonest on estuaries.

Ferruginous duck Recorded once at Towyn, Merioneth.

Long-tailed duck A rare visitor to estuaries and coast.

Velvet scoter A few off coast some winters. Also recorded Bala Lake.

Common scoter In large flocks off coast most winters, sometimes in summer.

Eider Occasional off coast, sometimes even in summer.

Hooded merganser Two of these American ducks were 'shot near Barmouth in 1864 by a clever right and left'(!).

Red-breasted merganser B After breeding in Wales (Anglesey) in 1954 for the first time on record, it has spread to several Merioneth estuaries and now breeds in some numbers.

Goosander Occasional visitor to estuaries and lakes.

Smew A rare visitor to estuaries.

Shelduck B Many breed on the estuaries. Numbers are lowest August-December when great majority are presumably on German coast, as ringing results indicate.

Greylag goose A few records only.

White-fronted goose Much the commonest goose, chiefly frequenting estuaries and coastal pastures. Often visits the upland bogs and lakes.

Bean goose A very few records, Merioneth coast.
Pink-footed goose A few most winters on the coast.
Brent goose A rather rare visitor to the coast.
Barnacle goose Rarely recorded.
Canada goose A very occasional visitor.
Mute swan B Widely but thinly distributed.
Whooper swan, Bewick's swan } Both visit lakes and estuaries autumn to spring, whooper being the commoner species.
Golden eagle Formerly bred around Snowdon but not for about 200 years. Recorded extremely rarely since then.
Buzzard B Widespread and fairly numerous especially in Merioneth. Ranges to mountains but is commonest in deciduously wooded valleys.
Rough-legged buzzard Rare passage and winter visitor.
Sparrowhawk B Fairly common in lowland and semi-upland country.
Kite Occasional visitor.
White-tailed eagle A few old records, mainly nineteenth century.
Honey buzzard Recorded once or twice last century.
Marsh harrier Rare autumn–winter visitor, mainly to estuarine marshes.
Hen-harrier A winter resident in very small numbers ranging widely from coast to moorlands. A pair has nested, or attempted to nest, the past few years.
Montagu's harrier A rare passage migrant which has occasionally bred, the most recent occasion being when 3 young were reared in 1951.
Osprey Has occasionally been seen on autumn passage on the coast.
Hobby A few nineteenth-century records.
Peregrine B Formerly nested on many crags but since the widespread use of chlorinated hydrocarbons in seed-dressings over the last few years the number of successfully breeding pairs has rapidly shrunk to practically none.
Merlin B Breeds sparingly mainly on heather moorland and heathery crags. Rather rare but more widespread in winter.
Kestrel B Widespread and in good numbers.
Red grouse B Widespread but not very abundant on most heather moors. Probably commonest in the drier east, on the Berwyn.
Black grouse B Fairly numerous and steadily increasing in range with the spread of conifer plantations.

Red-legged partridge Formerly introduced in several places but probably now absent everywhere. Persisted about Aberdovey till at least 1939.

Partridge B Much decreased this century and now only thinly but widely scattered in lowland and semi-upland districts.

Quail Rare spring and summer visitor. Likely to breed only in years when quails are generally numerous in Britain. Last century it nested abundantly near Towyn, Merioneth, in the great quail year of 1870.

Pheasant B Even where not preserved it holds its own well especially in reed-beds and rough cover near lowland bogs. Has spread into the hill-country in some of the forestry plantations.

Water rail Rare as a breeding species and is most often seen in autumn and winter.

Spotted crake A rare or seldom detected passage migrant and winter visitor.

Baillon's crake Very rare vagrant.

Corncrake Rare summer visitor which perhaps breeds here and there every year. Is most likely in semi-upland districts.

Moorhen B Not common. Breeds sparingly from sea-level to weedy moorland lakes.

Coot B Winters in fair numbers but few breed. Resorts to estuaries and harbours in very cold weather.

Oystercatcher B Breeds in the few little-disturbed sites now left on the Merioneth coast. Has bred inland at Trawsfynydd Reservoir.

Lapwing B Widespread but rather thinly distributed as a coast-to-moorland breeding species. Much commoner earlier this century.

Ringed plover* B Breeds along the Merioneth shore but is increasingly disturbed by holiday-makers.

Grey plover Frequent on passage and in winter, chiefly on estuaries.

Golden plover B In flocks, mainly on shore, autumn to spring. A very local breeding species on heather moors.

Dotterel There are several old records and some recent ones of dotterel on spring or autumn passage on mountain tops in Caernarvonshire and Merioneth.

Turnstone Passage migrant in small numbers, mainly estuaries.

Snipe B Breeds from sea-level to the moors.

Great snipe There exists an undated specimen shot at Dolgellau.

Jack snipe Infrequent autumn-spring visitor, mainly lowland, sometimes in hills.

* Little ringed plover. One at Mochras, 9 April 1966.

Woodcock B Thinly distributed as a breeding species. Most abundant in hard weather near the coast.

Curlew B Breeds in many bogs and fields from coast to semi-uplands. Abundant on coast autumn-spring.

Whimbrel Common passage migrant especially on coast.

Black-tailed godwit Coastal passage migrant in small numbers.

Bar-tailed godwit Common coastal passage migrant.

Green sandpiper Frequent on estuaries autumn-spring. Occasionally in the hills.

Wood sandpiper Rare passage migrant.

Common sandpiper BS A well-known summer visitor breeding along rivers and lake edges.

Redshank B Common on estuaries all year. Mainly breeds in coastal marshes but a very few nest occasionally in the hills.

Spotted redshank A not uncommon passage migrant on estuaries, especially in autumn. Increasing.

Greenshank Common coastal passage migrant July-October. Occasionally winters.

Knot Fairly common coastal passage migrant.

Purple sandpiper Rare on coast. Not recorded inland.

Little stint Rather rare coastal passage migrant.

Dunlin B Very common on coast autumn-spring. Breeds very sparingly on moorlands.

Curlew sandpiper Infrequent coastal passage migrant.

Ruff Infrequent coastal passage migrant.

Sanderling Common coastal passage migrant.

Avocet One seen near Mawddach estuary in 1901.

Grey phalarope Infrequent passage migrant usually in autumn.

Stone curlew 1 shot near Towyn 6 January 1903.

Arctic skua Rare passage migrant.

Great skua Rare passage migrant.

Pomarine skua Rare passage migrant.

Great black-backed gull Pairs are sometimes present at mountain lakes in the nesting season but there are no recent records of breeding. Regularly forages inland, feeding on carrion sheep.

Lesser black-backed gull A passage migrant, frequent on coast and hills.

Herring gull B Abundant and increasing on the coast and in the last few years has begun to breed at mountain lakes.

Common gull Abundant autumn to spring.

Glaucous gull Very few records.

Iceland gull One old record from Towyn.

Little gull Very occasional on the coast.

Black-headed gull B A few scattered nesting colonies at upland lakes and on coast. Common in winter.
Kittiwake Occasionally occurs, usually oiled or storm-driven.
Black tern Infrequent passage migrant.
Sooty tern One killed at Barmouth 17 August 1909.
Common tern Frequent on coastal passage.
Arctic tern Frequent on coastal passage.
Sandwich tern* Frequent on coastal passage.
Little tern B The only breeding tern, now reduced to one very small, much disturbed colony.
Razorbill Occasionally flying off-shore: frequent victims of oil.
Guillemot Occasionally flying off-shore: frequent victims of oil.
Little auk Rare storm victim.
Puffin Occasional oil victim.
Pallas's sandgrouse The first British record was of one shot near Tremadoc on 9 July 1859. Others were seen at Mochras, Merioneth, in 1888.
Stock dove B Rather thinly distributed up to the lower hills where it nests in old buildings, quarries and rock-crevices.
Wood-pigeon B Very common. Large autumn flocks when acorns abundant.
Turtle dove Scarce passage migrant.
Collared dove B This invader from the Continent has been breeding in the Park near Towyn, Merioneth, since 1963 and is likely to spread to other districts. Mainly near human habitations.
Cuckoo BS Common. Mainly parasitises meadow pipit. Ranges into the hills.
Great spotted cuckoo One found dead Aberdovey 1 April 1955.
Barn owl B Widely distributed up to the moorland region.
Little owl B A few mainly near the coast and probably decreasing.
Tawny owl B By far the commonest owl.
Long-eared owl B Very little is known about this elusive owl in Snowdonia. It is known to breed in the uplands in clumps of conifers and isolated hawthorns, and is best detected by song in late winter and early spring: a single hoot every two seconds.
Short-eared owl B Scarce nester mainly on wet moorlands, especially among recently planted conifers where it may become temporarily common if voles abundant. Thinly distributed in winter.
Nightjar BS Not uncommon in woodlands and on brackeny hillsides.

* Roseate tern One at Mochras 25 June 1966.

Swift BS Common in villages but ranging far over mountains after flying insects.

Kingfisher B Very local breeding species on lower reaches of rivers. More widespread autumn and winter on coast.

Hoopoe Rare passage migrant.

Green woodpecker B Widespread and fairly numerous. Visits open hillsides in search of ants.

Great spotted woodpecker B Plentiful in deciduous woodland. Less so in conifers.

Lesser spotted woodpecker B Rather scarce and mainly confined to lowland deciduous woods. Increasing the past few years.

Wryneck One shot near Barmouth 24 April 1900.

Woodlark B Mainly restricted to southern parts of the Park.

Skylark B Abundant nester lowlands and mountains. High ground largely deserted autumn and winter.

Swallow BS Common summer visitor and passage migrant.

House martin BS Rather local. Breeds in colonies in villages and on isolated houses sometimes in the hills. Occasionally in quarries.

Sand martin BS Local colonies in river banks and roadside gravel pits.

Golden oriole No definite record. Observers are advised to make sure bird is not a green woodpecker before claiming it as an oriole. (See p. 20.)

Raven B Widespread, nesting on crags and trees.

Carrion crow B Abundant, ranging into the mountain regions.

Hooded crow Rare winter visitor.

Rook B Common in the lowlands. Feeds on the hills in summer.

Jackdaw B Common. Some large breeding colonies such as at Tremadoc cliffs. Feeds on the hills in summer with rooks.

Magpie B Common up to the deciduous tree-line and some in high conifer plantations.

Jay B Common up to the deciduous tree-line and some in high conifer plantations.

Chough B Widely but thinly distributed, nesting in old quarries, mine-shafts and natural rock cracks. Commonest in north Merioneth. Flocks in autumn.

Great tit B Common up to the deciduous tree-line.

Blue tit B Common up to the deciduous tree-line.

Coal tit B Frequent in deciduous and coniferous woodland.

Marsh tit B Very local throughout the Park.

Willow tit B Widely distributed in the Park up to the highest level of deciduous woodland. Characteristic of thickets at the edges of bogs.

Long-tailed tit B Frequent in lowlands and also in upland conifer plantations.
Treecreeper B Frequent in lowlands and also in upland conifer plantations.
Nuthatch B Well established in most deciduous woodland especially in lowlands.
Wren B Common in many different habitats including screes and cliffs of high corries where it also winters.
Dipper B Widespread along streams and rivers well into the uplands. Recorded breeding up to 2,200 ft.
Mistle thrush B Common up to the limit of deciduous woodland.
Fieldfare Numerous autumn to spring especially in semi-uplands.
Redwing Numerous autumn to spring especially in semi-uplands.
Song thrush B Common from lowlands to upland conifers.
Ring ouzel BS Frequent summer resident on crags and screes in upland valleys and quarries. The song, a piping note four or five times repeated, is a characteristic sound of high rocky places. Also on margins of conifer plantations.
Blackbird B Common from lowlands to upland conifers.
Wheatear BS Summer resident widely distributed especially in open hill country with scattered rocks or walls.
Stonechat B Formerly well distributed in lowlands and semi-uplands but has decreased and is now mainly coastal.
Whinchat BS Frequent summer resident especially on wet moorland.
Redstart BS Highly characteristic of deciduous woodland especially on hillsides. Also in more open country with walls along the fringes of woodland.
Black redstart Infrequent passage migrant and winter visitor nearly always coastal. In winter usually in towns or villages.
Robin B Common from lowlands to upland conifers.
Grasshopper warbler BS Widely but thinly distributed in woodlands and marshes. Occasionally on fairly high moorlands.
Reed warbler There are old claims of it breeding and on passage but Snowdonia is far from its normal range.
Sedge warbler BS Locally plentiful in wet lowland habitats.
Blackcap BS Fairly common in lowland deciduous woodland. Rare in higher woodlands.
Garden warbler BS Characteristic especially of woodlands and thickets in semi-upland valleys. Also common among young moorland conifers.

Whitethroat BS Widely distributed from coastal cliff-tops to young moorland conifers.
Lesser whitethroat B? Rare summer visitor.
Dartford warbler One at Tremadoc May 1932.
Willow warbler BS Abundant from lowlands to young moorland conifers. Common on passage.
Chiffchaff BS Habitat much as blackcap but commoner. Abundant on passage.
Wood warbler BS Characteristic species of hillside oakwoods. Occasionally in conifers.
Goldcrest B Widespread in woodland especially conifers.
Spotted flycatcher BS Common in lowlands and extends thinly up wooded gorges to the semi-uplands.
Pied flycatcher BS Highly characteristic of hillside oakwoods.
Hedge sparrow B Common in scrub habitats and young moorland conifers.
Alpine accentor One seen on the Llanberis track up Snowdon 20 August 1870, one of the more northerly British records of this mainly south European species.
Meadow pipit B Commonest species of grassy places from coast to uplands. High ground largely deserted in winter.
Tree pipit BS Common especially on wooded hillsides.
Rock pipit B Resident along sea cliffs in south Merioneth.
Water pipit Three shot in the 1890's by the Glaslyn river near Portmadoc.
Pied wagtail B Common nester in lowlands and semi-uplands. Abundant autumn passage, but less frequent in winter.
White wagtail Frequent passage migrant.
Grey wagtail B The typical yellow-breasted wagtail of streamsides. Much less frequent in winter.
Yellow wagtail Infrequent passage migrant mainly coastal. Formerly bred by Bala Lake.
Blue-headed wagtail 2 shot on the Merioneth coast in the 1890's.
Waxwing Uncommon winter visitor usually seen eating berries along hedges or in gardens.
Great grey shrike Rare winter visitor that tends to haunt rather upland districts. May keep to one locality for several weeks.
Red-backed shrike Formerly common, now only a rare passage migrant.
Starling B Though abundant autumn to spring, comparatively few breed and nearly all are in towns or villages. Nests in trees are rather rare.
Hawfinch B Rare resident. Has bred lately near Dolgellau.

Greenfinch B Locally frequent especially in lowlands.
Goldfinch B Locally frequent especially in lowlands.
Siskin B Has increased and is now a widespread breeding species in tall conifer forests. Small numbers most winters.
Linnet B Common resident extending into moorland where there is gorse.
Twite Very little known. Past records of twites breeding or in large flocks are mostly suspect. Probably only a rare casual.
Redpoll B Widespread and increasing in forestry plantations especially Sitka spruce about 10 feet high, lowlands and moorlands.
Bullfinch B Nests in all sorts of thick cover including upland conifers but nowhere very common.
Crossbill Irregular visitor usually from early July onwards when in some years migrants arrive from Continent. Also breeds occasionally in conifers near Betws-y-coed and perhaps elsewhere.
Chaffinch B Nests in all hedgerow and woodland habitats. Large winter flocks.
Brambling Regular autumn-spring visitor usually in small numbers and roosting with chaffinches.
Yellowhammer B Common breeding species especially in gorse.
Corn bunting Formerly common resident especially on Merioneth coast. No record of nesting or even birds seen since about 1940.
Black-headed bunting There exists an undated specimen from near Towyn.
Cirl bunting Formerly near coast, no recent record.
Reed bunting B Locally common in lowlands ascending to moorland in a few places.
Snow bunting Scarce on passage and in winter, coasts and hills.
House sparrow B Towns and villages. Unusual at upland houses.
Tree sparrow Rare before the 1960's but is increasing as a winter visitor and is probably now breeding regularly in small numbers.

APPENDIX 3

BUTTERFLIES OF SNOWDONIA

(* = very local and scarce; ** = recorded once or twice only)

Large white
Small white
Green-veined white
Orange-tip
Pale clouded yellow**
Clouded yellow*
Brimstone
Silver-washed fritillary
High brown fritillary
Dark green fritillary
Pearl-bordered fritillary
Small pearl-bordered fritillary
Marsh fritillary
Comma
Small tortoiseshell
Large tortoiseshell**
Camberwell beauty**
Peacock
Painted lady
Purple emperor**
Red admiral
Speckled wood
Wall brown
Grayling
Meadow brown
Gatekeeper
Ringlet
Large heath
Small heath
Duke of Burgundy*
Small copper
Brown hairstreak*
Purple hairstreak
White-letter hairstreak*
Green hairstreak
Silver-studded blue**
Common blue
Holly blue
Grizzled skipper*
Dingy skipper
Small skipper*
Large skipper

APPENDIX 4

CONSERVATION BODIES

The Nature Conservancy,
Penrhos Road,
Bangor,
Caernarvonshire.

The Council for the Protection of Rural Wales,
Meifod,
Montgomeryshire.

The North Wales Naturalists' Trust,
Llys Gwynedd
Ffordd Gwynedd,
Bangor,
Caernarvonshire.

The West Wales Naturalists' Trust,
c/o The High School,
Dolgellau,
Merioneth.

GLOSSARY

Some of the words occurring in Snowdonia place-names

aber, mouth, confluence
aderyn (pl. *adar*), bird
afon, river, stream
allt, hillside
ar, upon
arth, hill
bach, small
bedd, grave
bedwen (pl. *bedw*), birch
benglog, skull
bere, kite, buzzard
betws, oratory
beudy, cowshed
blaen (pl. *blaenau*), head of valley
bod, dwelling
boeth, warm
bont, bridge
braich, arm
bran, cow
bras, prominent
bron, rounded hill
brwynog, rushy
bryn, hill
buarth, cattle fold
bwlch, pass
bychan, small
cader (*cadair*), seat
cae, field
caer, fort
cafn, trough
cam, crooked
canol, middle
capel, chapel
carn, *carnedd* (*pl. carneddau*), heap of stones, barrow, mountain
carreg, rock
caseg, mare
castell, castle
cefn, ridge
celli, copse
celyn, holly
cerrig, rocks
ceunant, ravine
cidwm, wolf
cigfran, raven
cil, nook
clogwyn, cliff
clyd, sheltered
coch, red
coed, woodland
congl, corner
cornel, corner
cors, bog
craig, rock
crib (pl. *cribau*), ridge
cribin, serrated ridge
croes, cross
crug (pl. *crugiau*), mound
cwm, valley, cirque
cwrt, court
cyfyng, narrow
cymer (pl. *cymerau*), confluence
darren, hill
ddol, meadow
ddu, black
ddwr, water
ddysgl, dish
deg, fair
derw, oaks
derwen, oak
deu (*dau*), two
diffwys, precipice
dinas, fort
dir, land

dol, meadow
domen, mound
draws, across
dref, hamlet, home
drum, ridge
drws, pass
du (pl. *duon*), black
dulas, dark stream
dwr, water
dwy, two
dyffryn, valley
dywarchen, turf
efail, smithy
eglwys, church
eira, snow
eithin, gorse
elen, *elain*, young deer
erw, acre
esgair, ridge
fach, small
fawn, peat
fawnog, peaty
fawr, large
fechan, small
fedwen (pl. *fedw*), birch
felin, mill
ffin, boundary
ffordd, road
ffos, ditch
ffridd, mountain pasture
ffrwd, waterfall
ffynnon, spring, well
figyn (*fign*), bog
filiast, greyhound
foel, bare hill
fraith, speckled, pied
fran, crow
fras, prominent
fron, rounded hill
fynach, monk
fynydd, mountain
gader (*gadair*), seat
gaer, fort
gafr, goat
gallt, hillside
gam, crooked
garn, rock
garth, hill
garw, rough
gelli, copse
gigfran, raven
gil, nook
glan, bank
glas (pl, *gleision*), green, blue
glyder (*gluder*), heap
glyn, glen
goch, red
goetre, woodland house
gors, bog
grach, scabby
graeanog, gravelly
graig, rock
gribin, serrated ridge
groes, cross
grug, heather
gwastad, level place
gwaun, moor
gwen, *gwyn*, white
gwern, marsh
gwernen (pl. *gwern*), alder
hafod (*hafoty*), summer dwelling
haul, sun
hebog, falcon
hen, old
hendre, winter dwelling
heol, road
hir (pl. *hirion*), long
hydd, stag
hyll, ugly
isaf, lowest
las, green, blue
llan, church, village
llanerch, glade
llawr, flat valley bottom
llech, slate
llechog, slaty
llechwedd, hillside
llefn, smooth
llethr, slope
llety, shelter, lodging

llithrig, slippery
lloer, moon
lluest, hut, bothy, summer dwelling
llwyd, grey
lwyn, grove
llyfn, smooth
llyn (pl. *llynnoedd* or *llynnau*), lake, pool in river
maen (pl. *meini*), stone
maes, field
march (pl. *meirch*), stallion
mawn, peat
mawnog, peaty
mawr, large
meillionen, clover
melin, mill
melyn, yellow
migyn (*mign*), bog
min, edge
moch, pigs
moel, bare hill
morfa, coastal marsh
mur (pl. *muriau*), wall
mwyn, mineral, ore
mynach, monk
mynydd, mountain
nadroedd, snakes
nant, stream, valley
newydd, new
odyn, kiln
oer, cold
ogof, *ogo*, cave
onnen (pl. *onn*), ash tree
pair, cauldron
pandy, fulling mill
pant, valley, hollow
pen, top
penmaen, rocky promontory
pennant, head of a glen
penrhyn, cape
pentre, *pentref*, village
perfedd, middle
person, parson
pistyll, waterfall
plas, mansion
poeth, warm
pont, bridge
pwll, pit, pool
'r, the, of the
rhaeadr, waterfall
rhedyn, bracken
rhiw, hill
rhos, marsh, moor
rhudd, red
rhwng, between
rhyd, ford
saeth (pl. *saethau*), arrow
sarn, road, especially a paved road
sych, dry
tai, houses
tair, three
tal, end
tan, under
tarren, hill
teg, fair
tir, land
tomen, mound
traeth, shore
traws, across
tref, hamlet, home
tri, three
troed, foot
tros, over
trum, ridge
trwyn, promontory
twll, hole
ty, house
tyddyn, small-holding
tywarchen, turf
uchaf, highest
uwch, above
waun, moor
wen, *wyn*, white
wrach (*gwrach*), witch
wyddfa (*gwyddfa*), mound, grave
y, the, of the

yn, in
ynys, island, riverside meadow
yr, the, of the
ysbyty, hospice
ysgol (pl. *ysgolion*), school, ladder
ysgubor, barn
ystrad, valley floor
ystum, bend in river

A SHORT BIBLIOGRAPHY

(omitting most of the many guide-books and tours)

APPERLEY, N. W. (1926), *A Hunting Diary*

BENOIT, PETER and RICHARDS, MARY (1963), *A Contribution to a Flora of Merioneth*

BIBLE, E. H. T. (1926–48), MS. diary

BOLAM, G. (1913), *Wild Life in Wales*

BOWEN, E. G. (ed.) (1957), *Wales*

BYNG, J. (1934), 'A Tour to North Wales' (1784) in *Torrington Diaries*, vol. 1

CAMDEN, W. (1695 and 1733), *Britannia*

CARR, H. R. C. and LISTER, G. A. (1948), *The Mountains of Snowdonia*

CARTER, P. W. (1955), 'Botanical Exploration in Merionethshire,' *The Merioneth Miscellany*

CATHRALL, W. (1828), *History of North Wales*, vol. ii

Council for the Protection of Rural Wales. Annual Reports

DAVIES, E. (ed.) (1957), *A Gazetteer of Welsh Place-Names*

DAVIES, WALTER (1797), *The Agriculture and Domestic Economy of North Wales*

DODD, A. H. (1951), *The Industrial Revolution in North Wales*, 2nd. ed.

EVANS, E. PRICE (1932), 'Cader Idris,' *J. Ecol.*

(1945), 'Cader Idris and Craig y Benglog.' *J. Ecol.*

EVANS, REV. J. (1800), *A Tour through part of North Wales*

FEARNSIDES, W. G. (1905), 'On the Geology of Arennig Fawr and Moel Llyfnant.' *Quart. Journ. Geol. Soc.*, vol. 61

Forestry Commission. *Forest Park Guide to Snowdonia*

FORREST, H. E. (1907), *The Vertebrate Fauna of North Wales*

(1919), *Handbook to Vertebrate Fauna of North Wales*

GRIFFITH, J. E. (1895), *Flora of Anglesey and Caernarvonshire*

GUNTHER, R. T. (1945), *Life and Letters of Edward Lhuyd*

HUGHES, R. E. (1958), 'Sheep population and environment in Snowdonia.' *J. Ecol.* 46, 169-90

(1960), 'The ecological approach to problems in land-use in upland Wales.' *Geography Dept. Memorandum* No. 3, University College of Wales, Aberystwyth

HUGHES, R. E. and NEILL, J. (1964), 'The Names of Crags in Snowdonia.' Part I. *Climbers' Club Journal* vol. xiv No. 2. New Series, No. 89, 172-184

HYDE, H. A. (1961), *Welsh Timber Trees*

HYDE, H. A. and WADE, A. E. (1957), *Welsh Flowering Plants*

(1962), *Welsh Ferns*

KILVERT, REV. F. (1944), Selections from *Diary*, 1870–9

LEWIS, S. (1833), *A Topographical Dictionary of Wales*

MATHESON, C. (1932), *Changes in the Fauna of Wales in Historic Times*

MAXWELL, SIR H. (1904), *British Fresh-water Fishes*

Nature Conservancy. Annual Reports

NICHOLSON, E. (1840), *Cambrian Traveller's Guide*

NICHOLSON, E. M. (1957), *Britain's Nature Reserves*

NORTH, F. J., CAMPBELL, B. and SCOTT, R. (1949), *Snowdonia*

NORTH, F. J. (1946), *The Slates of Wales*

PENNANT, T. (1784), *A Tour in Wales*

PRATT, ANNE (n.d.), *The Flowering Plants and Ferns of Great Britain*

ROBERTS, EVAN and ROBERTS, R. H. (1963), 'Plant Notes from South-East Caernarvonshire.' *Proc. B.S.B.I.*

Royal Commission on Ancient Monuments: (1921) *Merioneth*; (1956, 1960, and 1964) *Caernarvonshire*

SMITH, B. and GEORGE, T. N. (1961), *British Regional Geology: North Wales*

SMITH, S. G. (1948), 'The Butterflies and Moths of Cheshire, Flintshire, Denbighshire, Caernarvonshire, Anglesey and Merioneth.' Supplements in 1949 and 1950. (*Proceedings of the Chester Society of Natural Science, Literature and Art*)

Snowdonia Park Joint Advisory Committee. Annual Reports

STAPLEDON, R. G. (1936), *A Survey of the Agricultural and Waste Lands of Wales*

THOMAS, TREVOR M. (1961), *The Mineral Wealth of Wales and its Exploitation*

TURNER, D. and DILLWYN, L. W. (1805), *The Botanist's Guide*

WARD, F. (1931), *The Lakes of Wales*

WILLIAMS, H. (1927), 'The Geology of Snowdon.' *Quart. Journ. Geol. Soc.*, vol. 83

WILLIAMS, J. (1830), *Faunula Grustensis* (Fauna of Llanrwst)

WILSON, A. (1947–8), 'The Flora of a Portion of North-East Caernarvonshire.' *Northw. Nat*, 21 and 22

WOODHEAD, N. (1933), 'The Alpine Plants of the Snowdon Range.' *Bull. Alpine Garden Soc.*, vol. 2

The journal *Nature in Wales*, published jointly by the North Wales and West Wales Naturalists' Trusts, contains many informative papers and notes about the natural history of Snowdonia.

INDEX

Plants are listed in this index under both their scientific and their English names, with cross-references from one to the other. But to avoid duplication, page numbers are given with the scientific entries only. Thus the reader will find the references to Alder, for example, under the entry *Alnus glutinosa.*